AUTOMATION AND ROBOTICS IN THE TEXTILE AND APPAREL INDUSTRIES

TEXTILE SERIES

Howard L. Needles, Editor

FABRIC FORMING SYSTEMS
By *Peter Schwartz, Trever Rhodes* and *Mansour Mohamed*

TEXTILE IDENTIFICATION, CONSERVATION, AND PRESERVATION
By *Rosalie Rosso King*

TEXTILE MARKETING MANAGEMENT
By *Gordon A. Berkstresser III*

TEXTILE WET PROCESSES: Vol. I. Preparation of Fibers and Fabrics
By *Edward S. Olson*

TEXTILE FIBERS, DYES, FINISHES, AND PROCESSES: A Concise Guide
By *Howard L. Needles*

AUTOMATION AND ROBOTICS IN THE TEXTILE AND APPAREL INDUSTRIES
Edited by *Gordon A. Berkstresser III* and *David R. Buchanan*

Other Title

TEXTILE DYEING OPERATIONS: Chemistry, Equipment, Procedures, and Environmental Aspects
By *S.V. Kulkarni, C.D. Blackwell, A.L. Blackard, C.W. Stackhouse* and *M.W. Alexander*

AUTOMATION AND ROBOTICS IN THE TEXTILE AND APPAREL INDUSTRIES

Edited by

Gordon A. Berkstresser III
David R. Buchanan
North Carolina State University
Raleigh, North Carolina

np
NOYES PUBLICATIONS
Park Ridge, New Jersey, U.S.A.

Library of Congress Catalog Card Number: 86-5204
ISBN: 0-8155-1077-2
Printed in the United States

Published in the United States of America by
Noyes Publications
Mill Road, Park Ridge, New Jersey 07656

10 9 8 7 6 5 4 3 2 1

Library of Congress Cataloging-in-Publication Data

Automation and robotics in the textile and apparel
industries.

Bibliography: p.
Includes index.
1. Textile factories--Automation. 2. Clothing
factories--Automation. 3. Robots, Industrial
I. Berkstresser, Gordon A. II. Buchanan, David R.
TS1449.A9 1986 677 86-5204
ISBN 0-8155-1077-2

Preface

It is certainly challenging and perhaps presumptuous to create a book which deals with material in so formative a stage as is the subject matter of this volume. Exploring at the leading edges of new technologies and new human/techno systems must have its own intrinsic rewards, for the chance of meaningful extrinsic rewards for any individual involved in research in textiles/apparel automation/robotization is very slim.

The authors of the papers contained in this volume have labored long and hard to understand segments of the field. Many of them participated in a recent seminar at N.C. State University School of Textiles, and some of the articles herein are the result of that seminar. All of the authors continue to be active in research and all of us feel that this book is merely the first organized collection of knowledge about this admittedly specialized subject matter. Several million people in the U.S.A. work in the textile and apparel industries. Thousands of people are engaged in research and/or organized study about the industries. Economists and governmental leaders are also concerned about the future prospects of the textile and apparel industries. We expect, and our publisher supports us in this feeling, that this book, by providing a focus, will stimulate information exchanges, help generate a high level of interest in all parties affected, and may, we hope, bring some challenges to the ideas expressed herein.

There is no doubt that from the standpoints of national self-interests, economic pay-backs, and technological feasibility, the dawn of the age of automation and robotics in textiles and apparel is upon us.

How well each firm or nation succeeds in making the transition from highly labor intensive to highly capital intensive organization depends, in large part, on how well the managers of existing firms understand the forces at work and plan for a future which includes a massive change in the technologies and human systems. In 1968 Peter Drucker's book, *The Age of Discontinuity,* spoke to the fact that coming discontinuities–developments so different that they are unpredictable in current terms–make prediction of the future in many segments of modern life difficult if not impossible. Part of the changes taking place in the textile and apparel industries is evolving from historical bases. Moving pieces of fabric from one place to another by computer-controlled automatic overhead tram-ways and replacing human operators on sewing machines which stitch pockets onto shirt fronts are examples of such continuous evolution. But another part of the changes focuses on discontinuous, revolutionary technology where things like laser beams cut fabric instead of the scissors, knives, and other metal blades that have done so for thousands of years, and things like ultra-sonic waves provide seams in some fabrics instead of the needle and thread systems which have done so since the dawn of time.

It is the discontinuities, which are impossible to predict, which provide so much of the interest in the field. It is the existence of the discontinuities which has made this book so difficult to assemble.

The editors are in debt to the contributors of the chapters for their willingness to share their work; to George Narita, Executive Editor, Vice President of Noyes Publications for his cooperation, support, and patience; to Mrs. Patricia Knipper for her prompt and accurate typing of the manuscript; to their children for the understanding of missed family times; and to their wives, Patti Berkstresser and Sara Buchanan, for their understanding of the type of obsession which drives normally sane men to attempt a task of this nature.

Raleigh, North Carolina
March, 1986

Gordon A. Berkstresser III
David R. Buchanan

Contributors

Gordon A. Berkstresser III
North Carolina State University
Dept. of Textile Management
and Technology
Raleigh, North Carolina

George Blomquist
Parkdale Mills, Inc.
Lexington Division
Lexington, North Carolina

Julia Bonham
North Carolina State University
Division of University Studies
Raleigh, North Carolina

David R. Buchanan
North Carolina State University
Dept. of Textile Engineering
and Science
Raleigh, North Carolina

Augusto Carroli
American Savio
Charlotte, North Carolina

T. Waller George
North Carolina State University
Dept. of Textile Engineering
and Science
Raleigh, North Carolina

Wolfgang Giloi
IRI/Int. Robomation/Intelligence
Carlsbad, California

N. Alan Hunter
Celanese Fibers Operations
New York, New York

Henry Kelly
Office of Technology Assessment
Washington, D.C.

Edwin M. McPherson
North Carolina State University
Dept. of Textile Management
and Technology
Raleigh, North Carolina

Lukas Michel
Sulzer Ruti
Gebruder Sulzer AG
Ruti (Zurich), Switzerland

Jack Rebman
Lord Corporation
Thomas Lord Research
Laboratory
Cary, North Carolina

Wesley E. Snyder
North Carolina State University
Dept. of Electrical and Computer Engineering
Raleigh, North Carolina

James C. Solinsky
IRI/Int. Robomation/Intelligence
Carlsbad, California

John Strand
Amphion, Inc.
Ann Arbor, Michigan

Kazuo Takeuchi
Tokyo Keizai University
Kokubunjishi, Tokyo, Japan

Nicholas Williamson
University of North Carolina-Greensboro
School of Business
Greensboro, North Carolina

Contents and Subject Index

Overview of Robotics

W.E. Snyder
North Carolina State University

In this first chapter, Wes Snyder discusses the fundamentals of robotics, including robot geometry, the problems associated with knowing and controlling the position of end effectors, and the advantages to be gained from interfacing with sensors.

Dr. Wesley Snyder obtained his Ph.D. in Electrical Engineering from the University of Illinois, and is Associate Professor in the Department of Electrical and Computer Engineering of North Carolina State University. He is one of the founding members of Robotics International of the Society of Manufacturing Engineering, and has been active in robotics teaching and research for some 10 years. He also is the author of an undergraduate textbook on robotics.

Overview of Robotics*

Background

In 1921, the Czechoslovakian playwright Karel Capek wrote a play entitled R.U.R. (Rossum's Universal Robots). In the Slavic languages such as Czech or Polish, the word "robot" means worker. (It may come as a surprise to a visitor to Warsaw to see posters about organizations for robots.) Today, in these countries, the word has both meanings, human and mechanical.

Capek's story, in which the mechanical servants of man revolted, became the first of a series of stories in which the robot became the "bad guy", reflecting the disenchantment with technology which was prevalent in the years following World

*Chapter 1 of Industrial Robots, (Prentice-Hall), 1984. Adapted with permission.

War I. Later science fiction authors began to view the intelligent automaton in a more favorable light, again reflecting a more positive attitude toward technology by society in general. Another science fiction novel, "I Robot" by Isaac Asimov, published in 1950, probably best exemplifies this change.

In his novel, Asimov dealt not with the physical aspects of the machine but with the mental processes of a machine that might be required literally to interpret and obey contradictory comments. It is interesting to note that the development of Asimov's theme was concurrent with the emergence of the computer into the public sphere - for it is the computer that has enabled the existence of robots today, and the radical decrease in cost of computation is solely responsible for the economic success and popularity of industrial manipulators.

For centuries, mechanical automata have been made by people in their own image, but when computer technology invested those automata with computational capabilities, practical applications were at last possible.

Since the computer plays such a key role in robotics, the emphasis of this chapter will be on the computer and what it must do to make a set of sensors, actuators, and links into a robot.

Some Robotics Terms

As we use the word "robot" it has a very specific meaning. Therefore, to avoid confusion, we will now define some terms.

Manipulator

By far, the most popular type of anthropomorphic machine is the manipulator. It most closely resembles a human arm and hand, and we will often refer to it using those terms. Virtually every machine discussed in this chapter is of the one-arm, one-hand type. Some development work has been done on walking machines, but emulation of legs or other parts of the anatomy is still largely in the research stage and will not be discussed here. Thus, manipulator means arm, and we will use the terms almost interchangeably.

Teleoperator

A manipulator that operates in a tightly coupled mode with a human operator is called a teleoperator. The best-known example of teleoperators are the remotely controlled "hands" which run with a protective screen between them and their human operators. Such systems are specifically designed for handling radioactive materials. Today's teleoperators may have computers in their control loops to accomplish remote control or to enable the operator to sense the forces exerted by the manipulator. Such machines are theoretically capable of running autonomously, but they are not designed to operate without a human in the control loop. Therefore, the term "robot" does not apply.

Industrial Manipulator

Industrial manipulator is the term normally applied to the types of robots with which we will deal. Applications of such machines are not, of course, limited to industrial tasks, and applications in the commercial and agricultural sector exist. However, by far the majority of applications of robots to date have been in industry, primarily automotive manufacturing and metalworking.

Industrial manipulators vary widely in their geometry and sensing capabilities, and in their sophistication of control.

We will discuss this variety in the next sections.

Manipulator Geometries

Manipulators come in an assortment of shapes and sizes, from the Unimate PUMA with a payload of 3 kilograms to the huge manipulator on the space shuttle, which can move 500 kg (in zero G), and heavy-duty industrial machines with payloads of over 1 ton. However, manipulators may be categorized into a fairly small set of geometric organizations that we will discuss in this section along with the impact of physical geometry on computation.

Cartesian Geometries

Often referred to as gantry cranes, these robots have joints that move in rectangular orthogonal directions. These are perhaps the easiest to deal with in the computer, for a movement of x inches by the X actuator corresponds directly to a motion of x inches by the hand. Figure 1 shows the geometric organization of a Cartesian robot schematically; examples of such systems in the field are shown in Figure 2.

Cylindrical Geometries

In a manipulator using cylindrical geometry, the first three joints correspond directly to the three principal variables of a cylindrical coordinate system: θ (rotation), h (height), r (reach). The cylindrical geometry is illustrated in Figures 3 and 4. If the hand position is maintained in the (θ, h, r) coordinate system, then hand position is continuously known. Alternately, if the hand position is to be specified in (x, y, z) (as might happen if the robot is to coordinate with a conveyor), transformations must be performed relating the two coordinate systems, as illustrated in the

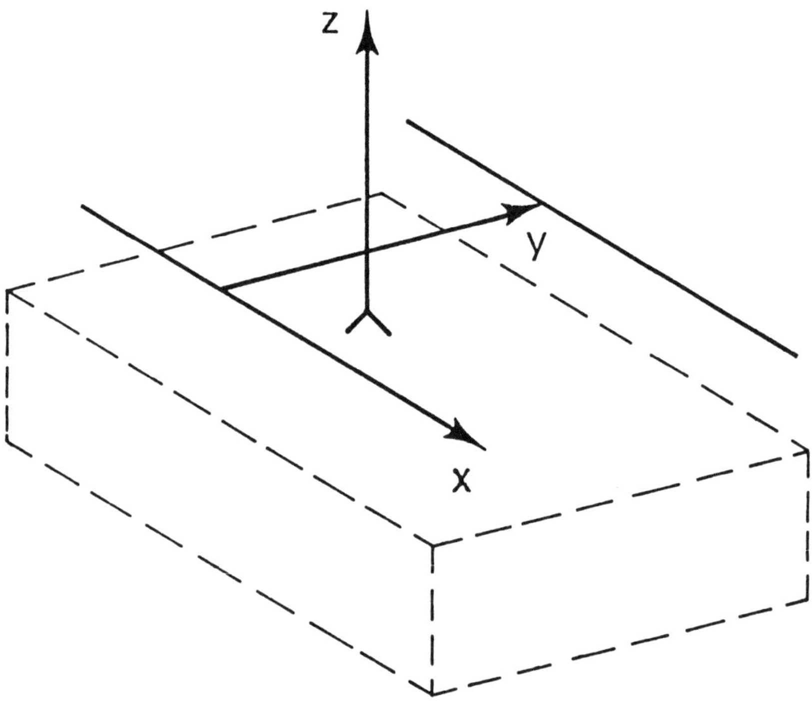

Figure 1. Rectangular coordinate system or "gantry crane" robot.

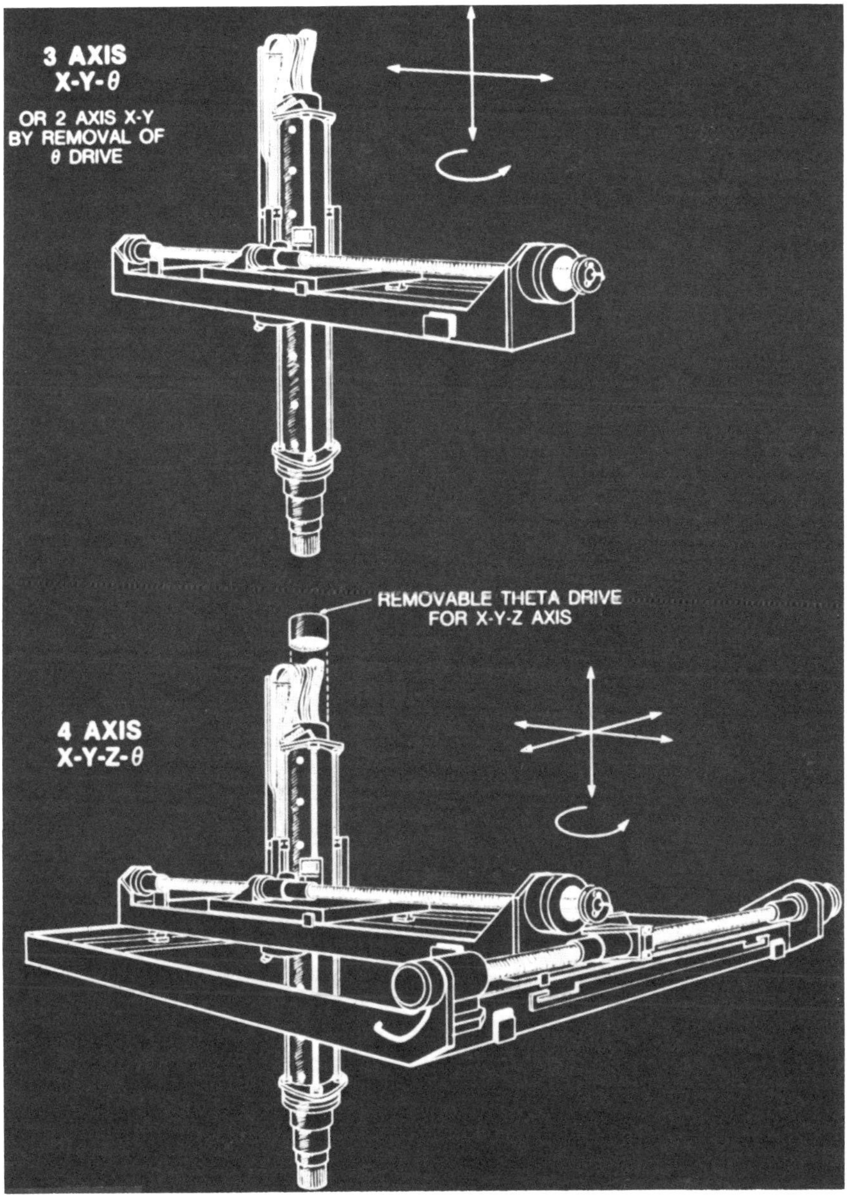

Figure 2. Gantry crane robots. Courtesy, General Electric.

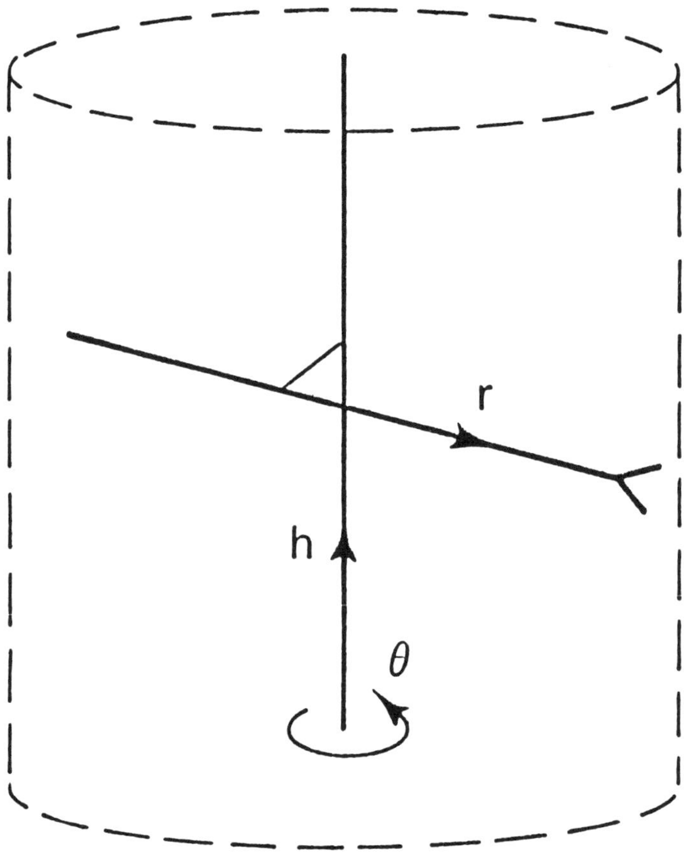

Figure 3. Kinematics of a cylindrical robot.

Figure 4. Cylindrical robot. Courtesy PRAB.

following example.

Example 1.1 Kinematics of a cylindrical robot

For the robot shown in Figure 3, determine the position of the hand in Cartesian coordinates (x, y, z) as a function of the joint coordinates (θ, h, r).

Solution: To solve this problem we will use geometric intuition.

The z axis is the same as the vertical axis of the robot.

Therefore, $z = h$

The x-y plane is perpendicular to the z axis and is, therefore, parallel to the plane in which θ rotates and r extends. Thus,

$$x = r \cos \theta$$

and

$$y = r \sin \theta$$

Spherical Geometries

As shown schematically in Figure 5, the first three joints of these robots correspond directly to the three principal variables of a spherical coordinate system: θ (rotation), ϕ (rotation perpendicular to the plane of θ), and r (reach or radius). If the hand position is maintained in this (θ, ϕ, r) coordinate system, then the hand position is known directly by reading the "joint variables". Alternately, if the hand position is to be specified as (x, y, z), then transformations must be performed to relate the hand position to the joint variables. These transformations are known as the kinematic transforms. Examples of robots using a spherical geometry are shown in Figure 6.

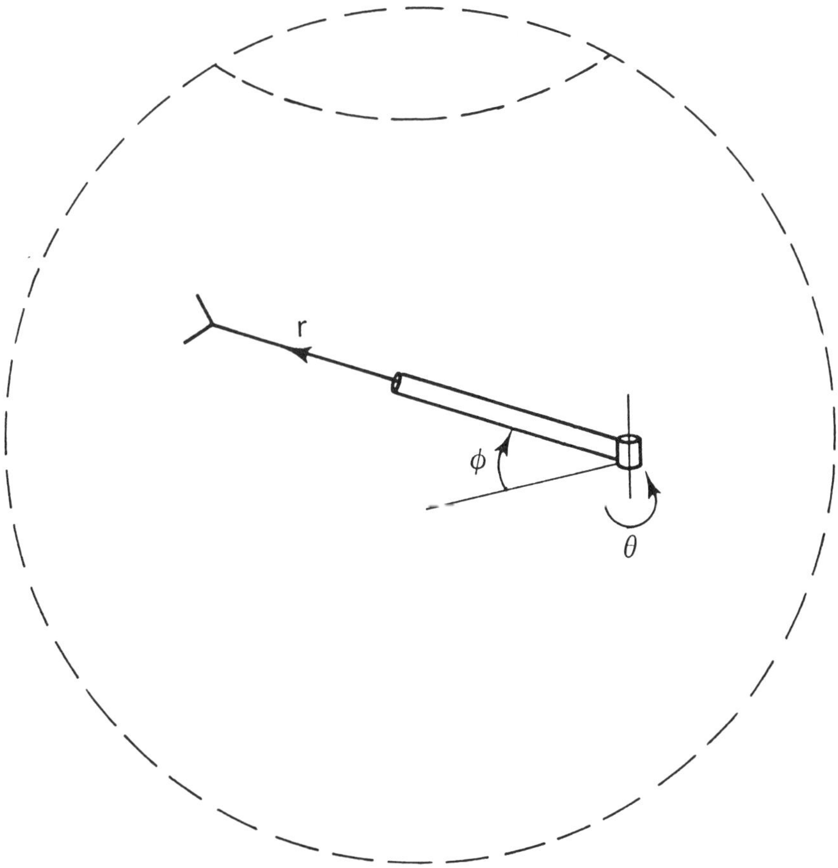

Figure 5. Spherical coordinate system.

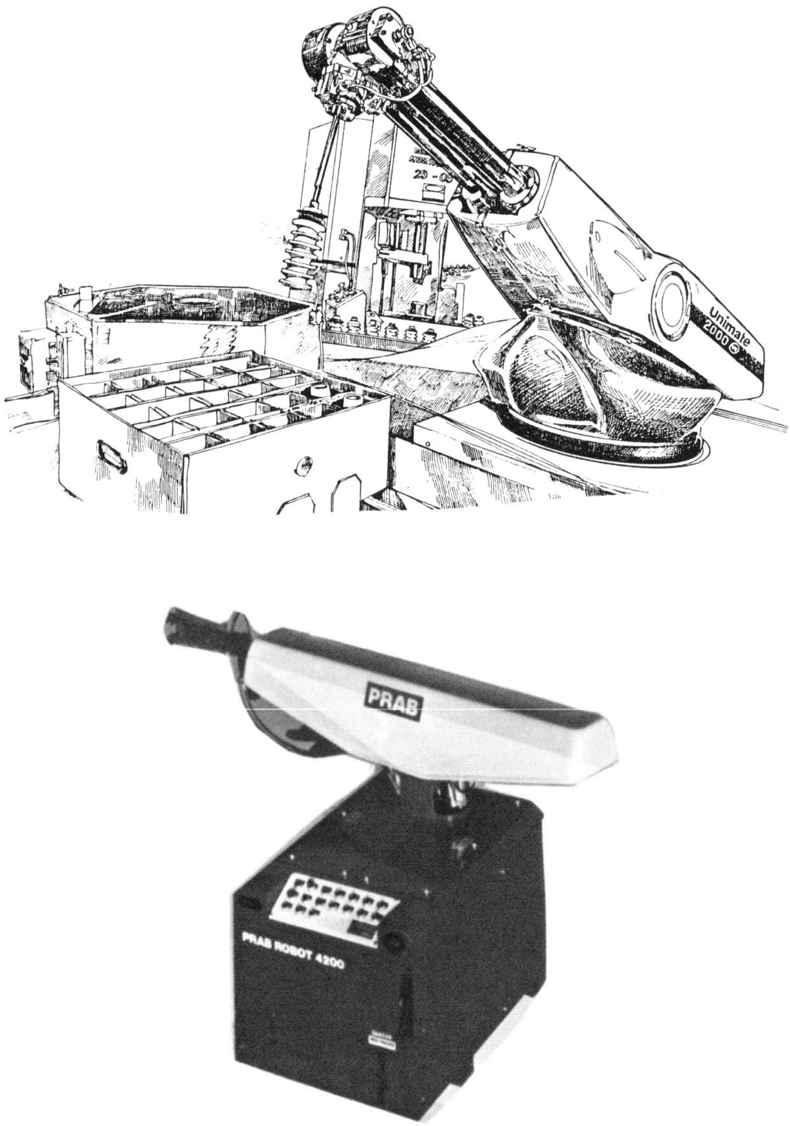

Figure 6. Spherical robots. Courtesy, Unimate & PRAB.

Example 1.2 Inverse kinematics of a spherical manipulator

For the spherical geometry of Figure 5, determine what the joint coordinates r, θ, ϕ must be to place the hand at a particular Cartesian position, x, y, z.

Solution: As in Example 1.1, we will use geometric intuition in solving this problem.

From the geometry shown in Figure 7 we determine the following relationships: we know that

$$y=r'\sin\theta,\ x=r'\cos\theta,\ z=r\sin\phi,\ r'=r\cos\phi$$

and

$$r^2=x^2+y^2+z^2$$

Furthermore, by dividing the first two equations, we find that

$$\frac{y}{x}=\tan\theta$$

$$\theta=\tan^{-1}\frac{y}{x}$$

Dividing the third equation by the fourth we find that

$$\frac{z}{r'}=\tan\phi \text{ and } \phi=\tan^{-1}\frac{z}{r'}$$

Thus we are able to find the three joint angles.

Articulated Geometries

Several robots now available emulate the human elbow, introducing a rather unusual set of joint variables, consisting of three rotations for the first three joints as shown in Figure 8. The transformations from the $(\theta_1, \theta_2, \theta_3)$ representation to (x, y, z) are even more complex. However, the articulated geometry is advantageous when the robot must reach over an obstacle. An example of a robot using articulated geometries is shown in Figure 9.

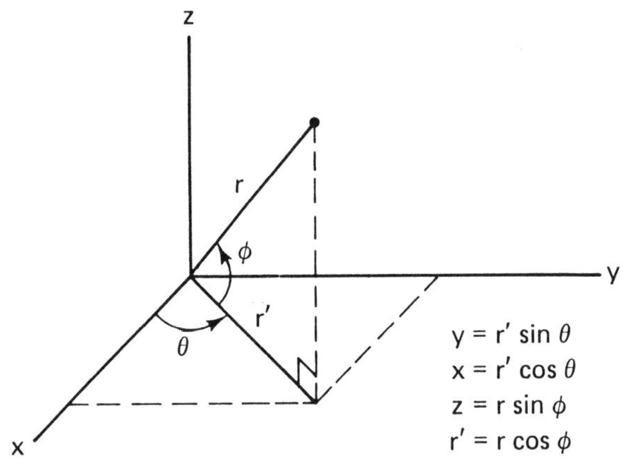

Figure 7. Inverse kinematics of a spherical manipulator.

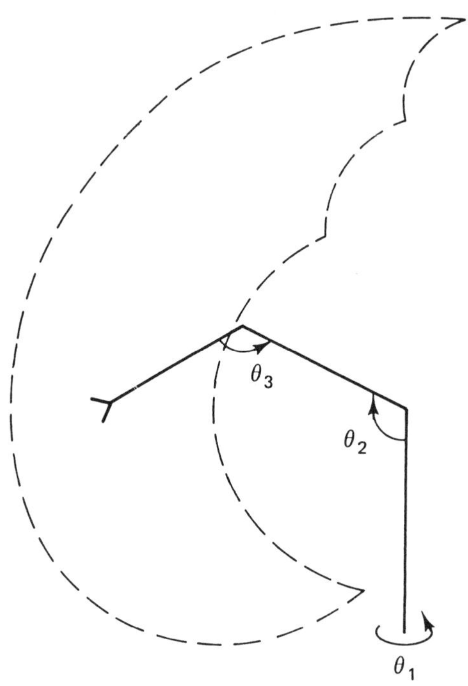

Figure 8. Articulated geometry.

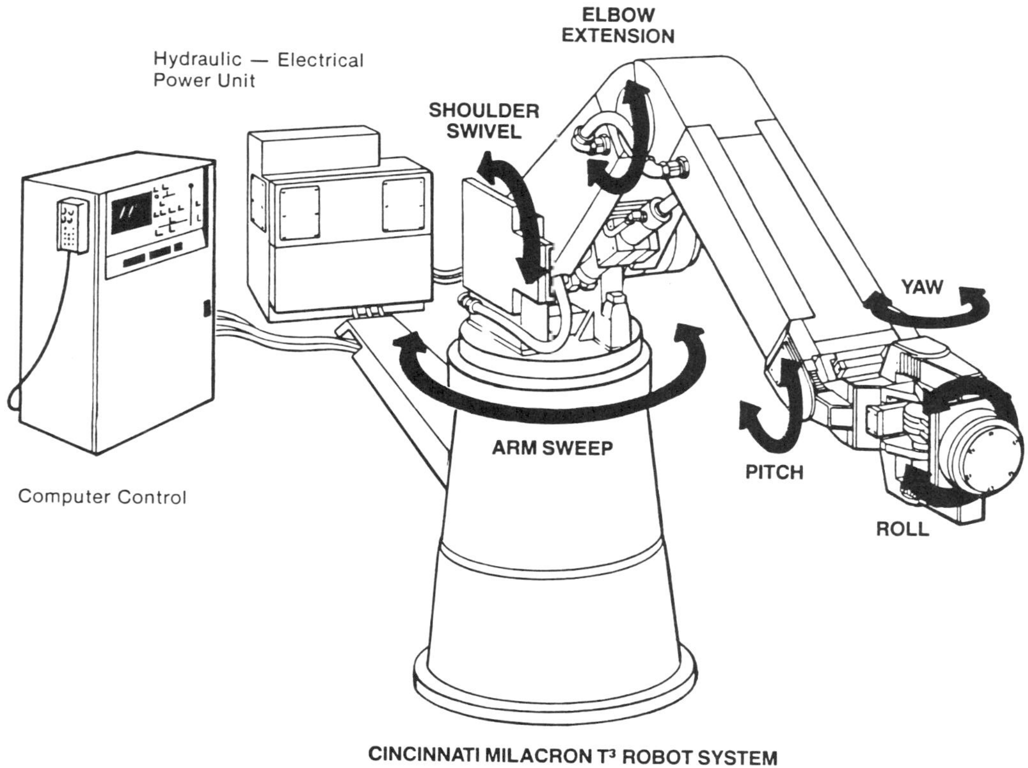

Figure 9. Articulated robot. Courtesy, Cincinnati Milacron.

Types Of Industrial Manipulators

Industrial robots may be categorized by their geometries, as described in the last section, or by their capabilities. In this section, we will discuss the capabilities of robots in terms of types of motions that can be accomplished. We will see that the degree of flexibility of control, like most other capabilities of robots, is directly related to the sophistication of the software within the controlling computer.

Limited Sequence: Mechanical Stop Control

The operation of mechanical stop robots is roughly analogous to that of manual typewriters and the early "programmable" sewing machines. In a manual typewriter, the tabs are set by mechanical stops. The act of pressing the tab-set key literally forces a small piece of steel out of the path of the carriage. The carriage then slaps up against this piece of metal and comes to an abrupt stop, generally in the correct place. Likewise, the older programmable sewing machines were programmed by inserting a complex cam into the machine. A motor slowly rotated the cam, and the eccentricity of the cam displaced the needle the appropriate amount.

Many of the simpler robots operate in exactly the same way. That is, an actuator (typically a pneumatic piston) moves a joint until that joint runs up against a mechanical stop. Programming of such robots is typically done with a screwdriver, although some flexibility may be built in by using several selectable stops on each axis. Some experts in the field claim that mechanically sequenced machines such as these are not robots.

Servo-Controlled Robots

If the control system is capable of detecting the instantaneous position of a joint, the actuator may be controlled in such a way that it can stop at any point along its path. It is this ability to stop at any point and not at just fixed mechanical stops that distinguishes, in many eyes, true robots from sequenced automation.

A servo-controlled robot can stop each of its axes at an arbitrary point. Therefore, its hand can reach any point in its working volume. Given this capability, the method of programming may still vary considerably. In point-to-point programming, the robot controller is given commands that amount to

Go to point A, stop.
Go to point B, stop.
Open gripper.
Go to point A, stop.
Go to point B, stop.
Go to point C, stop.
Close gripper.
And so on.

In between two points, say, A and B, no control is required over either the speed of the individual axes or the path of the hand. Since the individual axes may move at different rates, or have different (angular) distances to travel, one axis may reach its destination and stop long before another. The path of the hand in such motions is completely unpredictable.

Point-to-point control is commonly used in industrial environments in which the working volume is relatively empty (and the unpredictable hand motion is unlikely to cause damage)

and in which coordination with external moving objects such as conveyors is not required.

Continuous Path Control

In more complex work environments, the robot may be required to interact continuously with its environment. Examples include arc welding, spray painting, and performing operations along a moving conveyor. In such environments, the position and orientation of the hand must be exactly controlled all along its path. A variety of techniques exist to accomplish path control. The sophistication of the technique is dependent on such parameters as whether the path can be known in advance, as in spray painting, or whether it may vary during a motion, as in conveyor coordination. In all cases, a significant amount of data must be stored or computed.

Components of A Robot System

The components of a robot system could be discussed either from a physical point of view or from a systems point of view. Physically, we would divide the system into the robot, power system, and controller (computer). Likewise, the robot itself could be partitioned anthropomorphically into base, shoulder, elbow, wrist, gripper, and tool. Most of these terms require little explanation.

Consequently, in this section we will describe the components of a robot system from the point of view of information transfer. That is, what information or signal enters the component; what logical or arithmetic operation does the component perform; and what information or signal does the component produce? It is important to note that the same physical component may perform many different information processing operations (e.g., a central computer performs many different calculations on different data). Likewise, two physically separate components may perform identical information operations (e.g., the shoulder and elbow actuators both convert signals to motion in very similar ways).

From an information transfer point of view, a robot system appears as shown in Figure 10. In this model, we have ignored the physical and geometrical structure of the robot and are concerned only with the information flow in the system.

Actuator

Associated with each joint on the robot is an actuator which causes that joint to move. Typical actuators are electric motors and hydraulic cylinders. Typically, a robot system will contain six actuators, since six are required for full control of position and orientation. Many robot applications do not require this full flexibility, and consequently, robots are often built with five or fewer actuators.

Sensor

To control an actuator, the computer must have information regarding the position and possibly the velocity of the actuator. In this context, the term "position" refers to a displacement from some arbitrary zero reference point for that actuator. For example, in the case of a rotary actuator, "position" would really be angular position and be measured in radians.

Many types of sensors can provide indications of position and velocity. The various types of sensors require different mechanisms for interfacing to the computer. In addition, the industrial use of the manipulator requires that the interface be protected from the harsh electrical environment of the

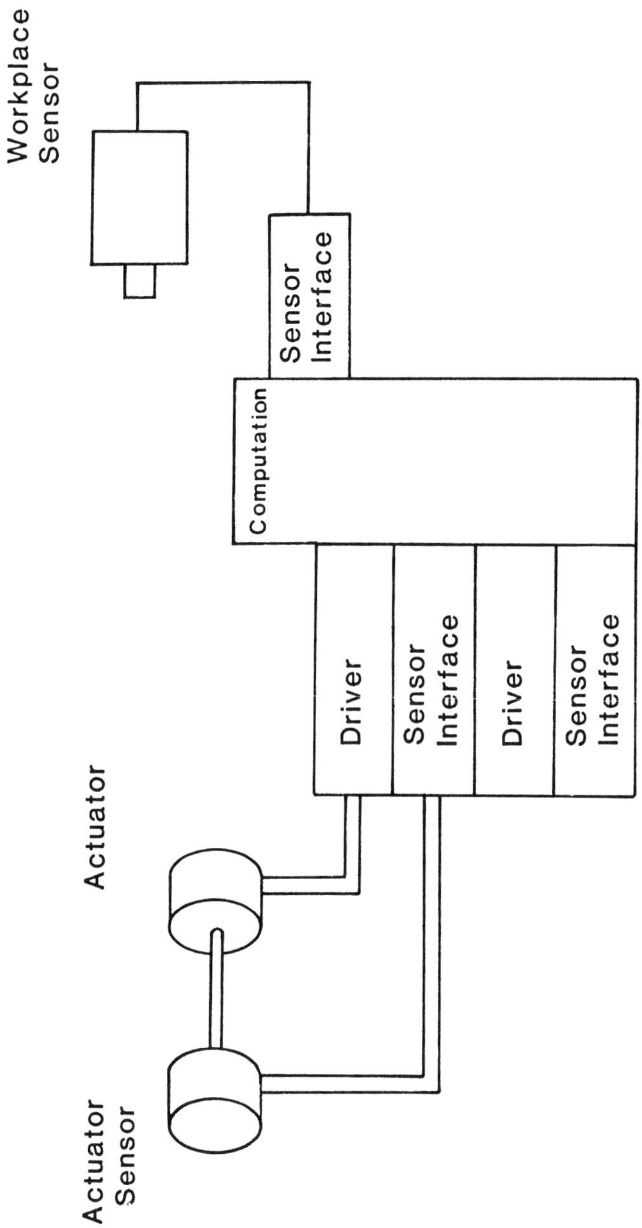

Figure 10. Components of a robot system.

factory. Sources of electrical noise such as arc welders and large motors can easily make a digital system useless unless care is taken in design and construction of the interface.

Computation

We could easily have labeled the computation module of Figure 1.10 "computer", as most of the functions to be described are typically performed by digital computers. However, many of the functions may be performed in dedicated custom hardware or in arrays or networks of computers. We will, thus, discuss the computational component as if it were a simple computer, recognizing that the need for realtime control may require special equipment and that some of this equipment may even be analog, although the current trend is toward fully digital systems.

One further note: We tend to avoid the use of the term micro-processor and simply say computer, although many current robot manufacturers use one or more microprocessors in their systems. The rapid advance of VLSI (Very Large Scale Integration) technology is producing levels of integration such that for all practical purposes, every computer is a microcomputer.

The computation component performs the following operations:

Servo. Given the current position and/or velocity of an actuator, determine the appropriate drive signal to move that actuator toward its desired position. This operation must be performed for *each* actuator.

Kinematics. Given the current state of the actuators (position and velocity), determine the current state of the gripper. Conversely, given a desired state of the hand, determine the desired state for each actuator.

Dynamics. Given knowledge of the loads on the arm (inertia, friction, gravity, acceleration), use this information to adjust the servo operation to achieve better performance.

Workplace sensor analysis. Given knowledge of the task to be performed (e.g., threading a nut), determine appropriate robot motion commands. This may include analyzing a TV picture of the workplace (vision) or measuring and compensating for forces applied at the hand (compliance).

In addition to these easily identified components, there are also supervisory operations such as path planning and operator interaction.

An Introduction to Machine Vision

W. Giloi and J.C. Solinsky
International Robomation/Intelligence, Inc.

Messrs. Giloi and Solinsky present the fundamentals of machine vision - the use of computer hardware and software to capture images and to extract information from them. In particular, they discuss the importance of obtaining the best possible image, the fundamental operations with which an image can be altered to allow its analysis, and strategies of obtaining information from those images.

Both authors are associated with International Robomation/ Intelligence, Incorporated, and have been involved in image analysis research and development for many years.

An Introduction to Machine Vision*

A NOTE ON THE NATURE OF MACHINE VISION

Image Analysis - The Key to Machine Vision

A vision system performs a variety of operations in image analysis, in order to carry out a pre-assigned task. In completing that task, two-dimensional images or even three-dimensional scenes are analyzed to extract useful information related to the task. The outcome of the task may be decisions

*Chapter 1 of A Primer on Machine Vision and the IRI Vision Systems. 1985. Adapted with permission.

as to the presence or absence of specific objects or the recognition and validation of certain attributes of pictorial objects such as size, location relative to other objects, completeness, flawlessness, and so on. While the human vision system evolves in a self-programming fashion from birth on, with a prior knowledge acquired by trial-and-error, computer vision requires individual programming of each particular task.

Thus, the first and foremost requirement in building an image analysis application program is to define the task to be performed by the system. From this definition evolves a hierarchy of procedural steps, spanning over a number of processing stages such as, to name the main ones:

- image acquisition
- feature enhancement
- image segmentation
- feature extraction
- recognition of objects and attributes of objects

Each of these steps involves a specific set of operations of the vision system and its associated peripherals. In very complex image analysis tasks, an additional processing step may be required pertaining to the recognition and understanding of the relationships between different objects. In the hierarchical, multi-stage image analysis process, two major phases can be distinguished, called the iconic domain and the symbolic domain of image analysis.

The Iconic Domain of Image Analysis

The iconic domain is characterized by the fact that operations are performed on the image, i.e., in the pixel space (icon is the Greek word for image). Functions in the iconic domain come in three major categories:

1. Feature enhancement functions are functions that transform images into new images. Their purpose is to enhance those features in the image that are significant for the image analysis task, while suppressing the insignificant features as much as possible. If we call the significant features the (useful) "signal" and the insignificant features the "clutter", then we can say that feature enhancement functions are filters that improve the "signal-to-clutter ratio" in an image. It must be emphasized at this point that the decision as to what is "signal" and what is "clutter" very much depends on the task to be performed.

2. Segmentation functions convert a gray scale image into a binary representation of the image, i.e., an image in which only two "levels of gray" exist, black and white. The reason for such "binarization" is that it usually facilitates the subsequent feature extraction step. Many industrial vision systems binarize the image acquired from the beginning; that is, such systems do not allow for the representation and processing of gray scale images. Consequently, vision systems of that type, called "binary vision systems", cannot perform feature enhancement functions. Many important computer vision tasks, however, cannot be carried out at all without elaborate enhancement filtering. Other tasks may be feasible on a binary system; however, the robustness (immunity against changes of the environmental conditions) of purely binary algorithms typically is much poorer than the robustness of the corresponding algorithms performed on gray scale systems where feature enhancement functions can be employed.

3. Feature extraction functions are operations applied on gray scale images or binary representations of gray scale images obtained first through the application of a segmentation function. The result of a feature extraction

function is not an image but a set of numerical values called a feature vector. Appropriate feature extraction functions will lead to feature vectors that are characteristic for the objects from which the features were extracted. As a general rule one can state that the feature extraction tasks become the easier the better the preceding feature enhancement filters work.

With feature extraction accomplished, the iconic domain of image analysis has been left, and the symbolic domain is entered. Figure 1 illustrates the steps of the iconic phase of image analysis.

Whereas the iconic domain of image analysis has reached a well-established state-of-the-art, the methodology of performing symbolic domain tasks is still in its infancy. It has become recognized, however, that complex scene understanding can be carried out only under elaborate use of *a priori* knowledge about the scene to be analyzed. It is safe to state that the complex image understanding tasks occurring in connection with robot vision, obstacle avoidance, process surveillance, product inspection, and quality control, and other sophisticated applications in computer integrated manufacturing (CIM) can be carried out in realtime only if the underlying vision system architecture combines very high speed execution of specific iconic domain functions with very powerful general purpose processing capabilities for the symbolic domain operations.

The Symbolic Domain of Image Analysis

The second major phase of the image analysis task deals with representations of the features extracted from an image rather than with the image itself. The feature representations in general are called symbols, and dealing with such symbols

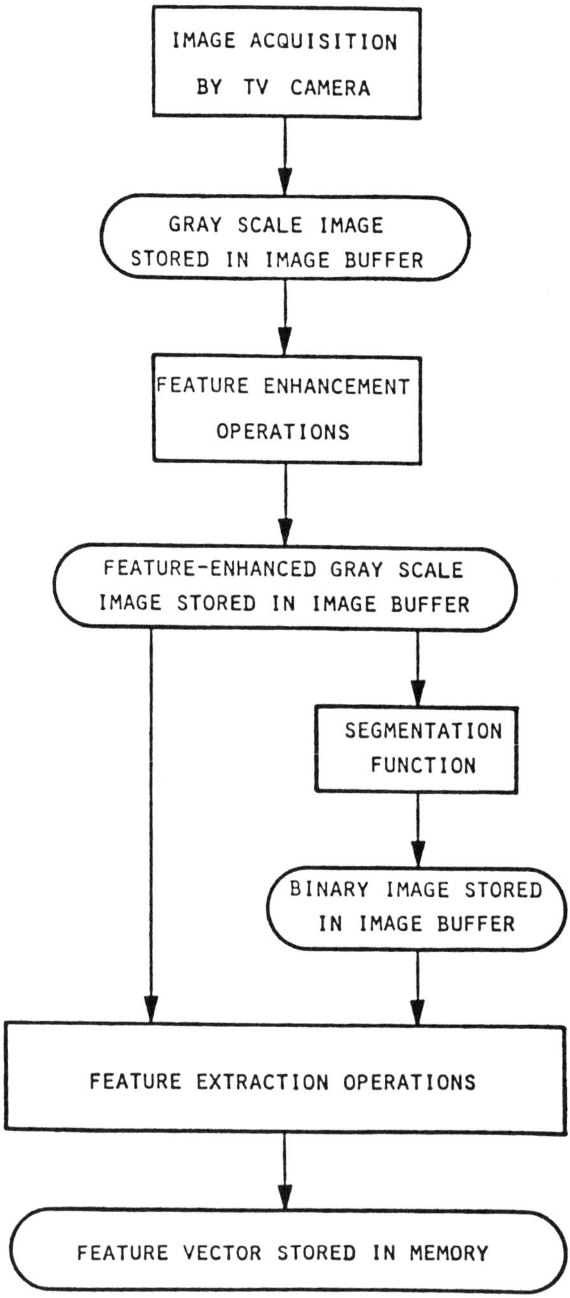

Figure 1. Steps in the iconic phase of image analysis.

therefore is called symbolic processing. In the symbolic domain of image analysis, the features extracted may be employed as input into a classification procedure, which usually is based upon either a syntactical or a statistical approach. Another form of classification is pattern matching. Here, certain features of the image are compared against a set of stored templates, and classification is based upon the best match found.

In general, the steps of the symbolic phase of image analysis may be the following:

- Recognition of objects and their attributes, respectively;
- Recognition of relationships between objects;
  Interpretation of relationships between objects ("scene understanding").

Usually, in typical applications of industrial image analysis, the iconic phase of image analysis accounts for 60 to 90 percent of the total number of operations to be performed. There is another idiosyncrasy to be found in the image analysis process that states that the higher one gets in the processing hierarchy, the less regularly structured are the data structures one has to deal with; however, the larger also is the degree of data reduction occurring at each stage.

To illustrate the above point, we compare two different tasks which both are representative of robot vision applications. The first task does not reach much beyond the iconic phase of image analysis, while the second task reaches deeply into the symbolic domain.

The first example considers the simple task of detecting the location and orientation of a single, isolated part. The part may be moving by the camera on a conveyor belt, so that a robot can pick it up. This means that, as a result of the image analysis procedure, an array of numbers containing the coordinates of the center of gravity and the angle of orientation of the part must be transferred to the robot controller. This enables the robot controller to move the arm and the gripper into the right position for picking up the object. The following sequence of activities of the vision system will render the desired result.

<u>Step 1</u>: Perform a point transformation to enhance contrast and suppress background noise.

<u>Step 2</u>: Threshold the gray scale image and calculate the first 3 moments of inertia.

<u>Step 3</u>: Calculate the angle of orientation from the values obtained in step 2.

<u>Step 4</u>: Communicate the values obtained to the robot controller.

The iconic realm is left at the end of step 2. From here on one deals with certain features, e.g., the first 3 moments of inertia, rather than with gray scale representations of binary representations of images. These operations can be performed with the standard preprogrammed functions of some commercial vision systems, including IRI Vision Systems.

The second example considers the very complex task of robot vision called the "bin picking problem". Here, a bin is filled with a number of different parts; and it is the task of the robot to pick a particular one. If the parts were neatly separated, the connectivity analysis program could be used to distinguish the various parts. This program performed the following steps.

<u>Step 1</u>: Perform some preprocessing for image enhancement,

e.g., a point transformation, and calculate the histogram.

Step 2: Based upon the histogram evaluation performed in step 1, threshold the image to obtain a binary representation.

Step 3: Perform a run-length encoding procedure that scans the image along horizontal scan lines and renders the coordinates of all transitions from 0 to 1 or from 1 to 0.

Step 4: Employ a "tree building" procedure that generates a tree structure, representing the point on the border lines of "blobs" and the relationships between them.

Step 5: Interpret the tree structure and identify the individual, isolated parts.

Here, the demarcation line between the iconic and the symbolic domain is reached at the end of step 3. In the IRI Vision Systems, run-length encoding is performed at very high speed by the array processor, while tree building must be done by the host computer.

If the parts in the bin overlap -- and this in practice may often be the case -- then the bin picking problem is rather complex. First of all, the solution requires edge information, which is not obtainable from a binary image. However, edge extraction alone will not solve the problem, as the edges can belong to many different, overlapping objects in the bin.

Without the use of *a priori* knowledge about the nature of the "scene" that can be expected to be found in the bin, for all practical purposes it is impossible to solve the bin picking problem. The use of *a priori* knowledge, though, may require the employment of sophisticated methods of artificial intelligence such as heuristic search procedures. It may even necessitate the help of an "expert system" to acquire the *a* priori knowledge needed in the first place. All these highly sophisticated techniques, such as heuristic searches, knowledge acquisition, etc., reside totally in the symbolic processing domain.

Consequently, the higher levels of symbolic processing require the flexibility of a conventional, "von Neumann" type computer, while the iconic stages of image analysis allow for the use of very fast array processors. Such a fast array processor can be realized most cost-effectively in the form of a high-speed, multi-staged pipeline or systolic array, operating in the "SIMD mode" (SIMD: Single Instruction-Multiple Data).

All these considerations led to the design of the IRI Vision Systems, in which a conventional "host computer" is supported by a number of "co-processors", i.e., fast array processors, which provide the computing power, needed during the iconic phase of image analysis, to perform industrial inspection tasks in realtime.

TASK DEFINITION

General Considerations

Defining the task in image analysis usually is the most crucial part of the problem. A precise formulation of the task at hand is the first step in using a vision system. Whereas humans can readily view an image, e.g., a television picture, and immediately point out the features on the image that are related to the task, a machine does not have such an innate capability. Since, in computer vision there is no "little man" inside the box viewing a monitor, machine intelligence must be provided instead by the application software. Consequently, the task definition becomes a responsibility of the application programmer. On the other hand, it is quite useful for the programmer to use his own "vision system" to evaluate the steps in the image analysis procedure s(he) is developing.

For example, identifying player number sixty-four on the Monday night football wide-screen television is trivial for the avid viewer. However, think for a moment what the actual steps are. Depending on the specifics, the following steps are quite likely:

1. Find the football player "signal" amidst the "clutter" of cheerleaders, spectators, and other football field activities.
2. Find the backside of the jerseys worn by the player.
3. Find the numbers on each jersey.
4. Identify the numbers from the set of possible integer values from 0 to 99.

The steps of the hierarchical image analysis procedure require considerable real-time processing capabilities and are quite comparable to the typical computer vision tasks in the industrial work place. Assuming one has more control over the images acquired, let's redefine the task, where this control has been exercised.

The new task is now to find the jersey number 64, when the three "best" players are lined up on the bench, and the camera is controlled to separately examine each backside. Since now the possible choice of numbers has been reduced to only three (the "best" players being a predefined set of numbers containing 64), one just has to compare the viewed numbers, which now occupy most of the image view, from a choice of three.

The procedure is similar for industrial computer vision tasks. The task definition includes controlled image acquisition in order to simplify the overall goal. This includes the control of the environmental conditions under which the image is acquired, especially the lighting conditions. In many tasks of industrial machine vision, providing the appropriate lighting often is the key to successful application of machine vision.

Image Acquisition

As discussed in the previous section, the image to be acquired and subsequently processed is related to the task definition. Therefore, the image must contain the appropriate scene assumed in the task definition. In general, the vision systems acquire an image through a television camera.

The scene of which images shall be acquired must be highlighted with appropriate lighting, and the appropriate lenses and optical filters should be employed to emphasize the image features related to the task, the "signal", and to suppress as much as possible the image features not related to the task, the "clutter". The image will contain also random features not very well defined, such as haze, dirty components, or dust. The random features are best described by statistical properties, i.e., are considered as "noise". Optimizing the signal to clutter ratio before acquiring the image by optimized lighting, etc. is well worth the effort.

The image acquired must find an appropriate digital representation in the vision system's memory in order to be processed by it.

The continuous tone image picked up by a black-and-white camera is composed of various levels of gray spanning over the entire gray scale ranging from black to white. The camera scans the image by measuring the gray levels along a raster of scan lines (Figure 2) and converts these measurements into the amplitude of an analog signal.

A digital computer, on the other hand, can process only discrete numbers. Consequently, the analog camera signal must be digitized before its values can be stored in the vision system's memory. Digitizing the camera signal means two things:

- The signal must be sampled in time, i.e., the continuous signal representation of the image is replaced by a sequence of discrete values taken at discrete instants of time. Usually, these time instants, called sampling points, are equally spaced in time; and the number of samples taken per time unit is known as the sampling rate.

- The continuous tone gray levels of the samples taken must be discretized; i.e., the gray scale is represented by an interval of discrete numbers.

This raises two questions, namely:

1. How many samples should be taken?
2. How many levels of gray should be distinguished?

The original image presents itself as a continuous distribution of gray levels in a two-dimensional space. The number of samples taken determines the spatial resolution of the digital representation of the image acquired.

The decision as to what spatial resolution to provide in a computer vision system is a trade-off between contradicting demands, viz.:

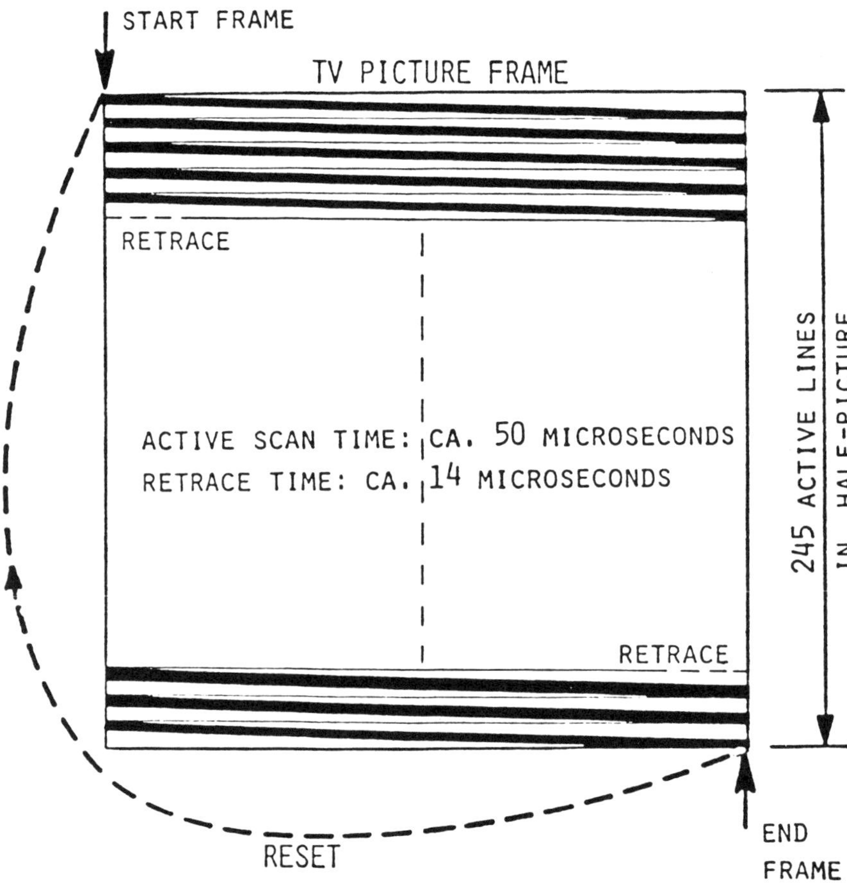

Figure 2. RS170 TV raster scan.

- The demand for high-definition image representations, in which no vital information of the original image is lost;

- The demand for high processing speed;

- The demand for maximal cost-effectiveness.

The finer the resolution of the digitized image is, the better is the original image approximated. On the other hand, the execution time of an image processing algorithm, in a first approximation, is proportional to the square of the number of pixels to be processed. Doubling the resolution means that 4 times the number of pixels are involved in the processing of the image. Consequently, either execution time increases by a factor of 4, or much more hardware must be spent in order to increase the processing speed by that same factor. Moreover, 4 times the memory capacity is needed to store the digitized images.

The question of what spatial resolution should be provided is answered by the sampling theorem. This theorem states that the sampling rate must be at least twice as high as is the highest spatial frequency in the image, in order to avoid a distorting effect known as aliasing.

Sampling a continuous signal has the effect that its frequency spectrum (its Fourier transform) is not a single spectral function (as is the spectrum of the continuous signal) but a periodic repetition of the spectrum of the original continuous signal. This periodic repetition is called the aliasing of the original spectrum.

Aliasing does not matter as long as the individual spectral functions in the aliased spectrum are well-enough separated from each other in the frequency domain. In this case, the higher spectra can be suppressed by a low-pass filter, leaving only the original spectrum, thus restoring the original signal. Given a sampling rate, there is a maximum frequency known as the Nyquist frequency that must not be exceeded by the signal in order to render the aliased spectra separable.

If the Nyquist frequency is exceeded, however, the components of the aliased spectrum overlap and, therefore, cannot be separated by any filter. The result is a distorted image.

There exist 2 possibilities to avoid aliasing:

- Prior to sampling, the original image is low-pass filtered, so that all spatial frequencies in the image that would exceed the Nyquist frequency are suppressed.

- The sampling rate simply is made high enough to obtain a Nyquist frequency higher than the highest frequency components the camera can deliver. Taking 256 samples per TV scan line usually is sufficient to avoid any serious aliasing problems.

In industrial image analysis it has become somewhat of a standard to use a spatial resolution ranging from 256x256 to 512x512 samples. In many cases, the lower resolution is perfectly adequate for the application to be performed and, consequently, cost-effectiveness of the system is maximized. Moreover, until recently, there existed hardly any solid state cameras with a resolution higher than 256x256.

However, there exist application areas (gauging, certain inspection tasks) where a resolution of 512x512 is at least desirable, if not required. Since solid state cameras with such higher resolution have begun to materialize, 512x512 resolution is about to become a second standard in industrial image analysis systems.

When discussing the question of how many levels of gray should be distinguished, one has two facts to go by:

- The human eye at best can distinguish 64 levels of gray.
- Most cameras cannot do much better; in the best case a camera may be able to distinguish about 100 gray levels.

However, one also must consider that it is in the nature of the A/D (analog-to-digital) converter, employed to convert the analog camera signal into a sequence of discrete numbers, to render an output the least significant bit of which is not very well defined. For all these reasons, no less that 8 bits should be used to represent the gray levels of the pixels sampled.

For further processing, the samples obtained from an image are stored in an <u>image buffer memory</u> in the form of a two-dimensional array, called the <u>gray level matrix</u>. These elements are called the <u>picture elements</u> or, in short, <u>pixels</u> of the discretized representation of the image. Their arrangement is such that the pixel matrix is isomorphic to the image it represents; i.e., the inter-pixel relationships of the original image are preserved. The stored framed then can be operated upon in order to extract and identify significant features of the image. Figure 3 illustrates the manner in which the pixels of a gray level matrix are indexed.

To understand the boundary conditions under which an image

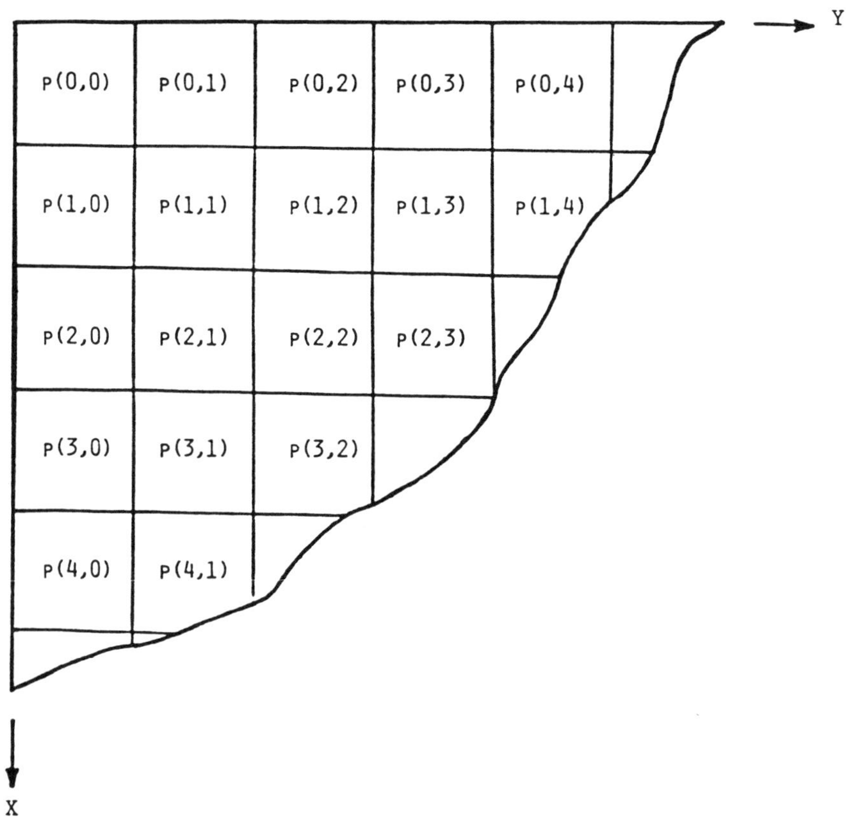

Figure 3. Indexing of the pixels of a gray level matrix.

can be acquired by a TV camera, let us take a look at the RS170 television standard.

According to the RS170 standard, a television picture consists of two interlaced half-pictures of 525/2 = $262\frac{1}{2}$ lines each. In vision systems with 256x256 pixel resolution, the interlace feature of the TV standard usually is not utilized, for the resolution of the half-picture is adequate. Figure 2 illustrates the raster scan scheme by which a half-picture is scanned.

The return of the scanning spot from left to right and from bottom to top of a frame consumes time, which therefore cannot be utilized for carrying image information. About 20% of the total scan line time of approximately 64 microseconds is needed for the scan spot <u>retrace</u> to the next line, and 35 out of the 525 lines are blanked out to reset the scanner from the end position to the start position of the next frame.

For a 256x256 pixel vision system, complying with the RS170 camera standard is highly desirable, since it allows for the use of cost-effective standard cameras. In view of the discussion above, however, there are two consequences for such an advantageous choice.

First, the 256 pixels on a scan line must be sampled during the "active portion" of a scan line, which extends over about 50 microseconds, resulting in a sample rate of over 5 million samples per second.

Second, since a half-picture has only 245 active scan lines, the resolution of 256 pixels in vertical direction can not totally be reached. Consequently, in order to obtain an even number, 240 raster lines are scanned and stored in the frame buffer memory. Note that in the European (CCITT) standard a half-picture has 312½ scan lines. This leaves about 290 active lines in the camera signal, so that it is in this case no problem to obtain the full 256x256 resolution.

Feature Enhancement

The major purpose of the feature enhancement step is to improve the ratio of useful information, the "signal", to the information one is not interested in, that is, "clutter" and "noise".

The signal-to-noise ratio is improved if the noise can be filtered out without changing the signal significantly. This normally is a fairly straightforward task, since signal and noise usually differ significantly in certain global characteristics. For example, noise may appear in the image in the form of isolated, bright pixels, which thus can readily be recognized. In the Fourier domain (spatial frequency domain), noise is characterized by high spatial frequencies, thus contrasting to the signal which may contain only lower frequencies.

The separation of signal and clutter is more difficult, for the global characteristics such as gray level distribution or spatial frequency distribution may be quite similar. Consequently, the differences may be found here only by looking for specific relationships between individual pixels. For example, if one is trying to find edges, the signal property now is the connectedness of a chain of adjacent pixels. If one is looking for straight lines, the connected pixels, moreover, must be colinear.

In computer vision, the purpose of <u>feature enhancement</u>, as opposed to <u>image enhancement</u>, is <u>not</u> to improve the quality of the image for the human viewer (e.g., by deblurring it, filtering out noise, enhancing contrasts, etc.) but to render the features better recognizable by the subsequent feature extraction operation. This leads us to a general definition of a "feature enhancement filter".

> A feature enhancement filter is a transformation of a gray level image into a new gray level image such that certain features of the original image are represented in the new image by more characteristic gray scale patterns.

For example, let us assume we have a "scene" in which we expect to find rectangular boxes whose sides are more or less parallel to the axes of the x-y coordinate system. This is what we call "<u>a priori knowledge</u>" of the scene. The task shall be to find the corners of the boxes in order to determine the location of each box. The 4 corner forms to be detected shall be denoted by ┌ , └ , ┘ , and ┐.

The first step in this procedure consists in applying a "corner detection filter". Such a filter is provded, for instance, by a 3x3 convolution called the <u>Prewitt operator</u>. For each of the 4 corners, there is a specific set of convolution coefficients, viz.

┘:

$$C1 = \begin{matrix} 1 & 1 & 1 \\ 1 & -2 & -1 \\ 1 & -1 & -1 \end{matrix}$$

┐:

$$C2 = \begin{matrix} 1 & -1 & -1 \\ 1 & -2 & -1 \\ 1 & 1 & 1 \end{matrix}$$

┌:

$$C3 = \begin{matrix} -1 & -1 & 1 \\ -1 & -2 & 1 \\ 1 & 1 & 1 \end{matrix}$$

└:

$$C4 = \begin{matrix} 1 & 1 & 1 \\ -1 & -2 & 1 \\ -1 & -1 & 1 \end{matrix}$$

Corners are detected by convolving the image P with each of the 4 corners detection functions, C1 to C4:

$$Q = \sum_{m=-1}^{+1} \sum_{n=-1}^{+1} Ck(i-m,j-n)xp(i,j); \qquad 1 \leq k \leq 4$$

A corner is indicated by the condition

$$Q \geq 4$$

The task shall be to detect the location of the upper left corner of the rectangles superimposed to the gray scale image. To this end, the following steps are taken:

1. Scale down the gray levels by $\frac{1}{4}$ in order to avoid overflow;
2. Apply appropriate Prewitt operator;
3. Threshold result to suppress all gray values except the 2 peaks;
4. Apply region growing operator to make result more visible.

The Prewitt operator, as we have seen, convolves every 3x3 neighborhood of the gray scale image with a template for each of the 4 corners. Actually, in mathematical terms the operation performed is a "non-normalized correlation function". A process of this type is called template matching and the templates employed are also called matched filters. Template matching is one of the most effective methods of feature enhancement. Therefore, we shall give this method some more consideration.

A Note on Template Matching

One of the most frequent tasks in image analysis is to find out whether an image contains a particular object. If present in the image, the exact location of the object in the image is asked for. Another problem occurring frequently is to have two images showing the same scene at different instants in time or under different angles of view in a three-dimensional space. Here, the task may be to recognize features of the first image in the second image ("scene matching" or "image registration"). In general, problems of that type may be solved by a template matching procedure. In the case of object detection, for example, the object searched for is the template that is compared to the image under investigation.

Given: a gray scale matrix P of resolution nxn, the image, and a second gray scale matrix of resolution mxm, the template, with $m<n$. In order to find out whether the image contains the template, one can place successively the template origin onto all given pixel positions (i,j), measuring in each position the degree of the match between the image and the template. The degree of match can be measured by calculating the cross-correlation between image and template in each possible position. A match is assumed to exist if the maximum of the cross-correlation function exceeds a given threshold value; otherwise, there is no match. To support this important task of template matching, the set of co-processor functions in the IRI Vision Systems comprises the fast execution of the cross-correlation function.

Image Segmentation (Binarization)

Image segmentation is the process of converting a gray level image into a binary representation of the significant features (the "signal"). Prior to feature extraction, certain feature enhancement routines may be applied to the image to emphasize the features to be extracted and, thereby, improve the "signal-to-clutter ratio". The features extracted either may become the objects of subsequent image understanding routines or they may be numerical measurements that constitute an end in itself.

A general view of features is that of a collection of adjacent pixels that have certain attributes in common. The general approach to the detection of specified features is through connectivity analysis, i.e., through a procedure that finds such a set of pixels with common attributes. The typical approach is to accomplish this in two steps, the first step consisting in a filtering procedure, by which the features of interest are emphasized, and the second step consisting in a detection algorithm for the enhanced features.

The generic term for a process that separates the features of interest from the "clutter" is image segmentation. Image segmentation, in general, denotes the process of decomposing an image into simpler components which, hopefully, provide a good approximation of the features one is interested in. There exist two "natural" ways of segmenting an image, viz.

- segmentation into homogeneous regions
- segmentation into contours.

Both approaches lead to similar results in different ways. The homogeneous contour approach utilizes the property of an image that neighboring pixels satisfy a given homogeneity criterion, e.g., that the difference in their gray levels will not exceed a given threshold. Segmentation methods such as "region growing", "merging", "blob labeling" all fall into that category. In the contouring approach on the other hand, rather than determining regions, one tries to find the contours, i.e., the borderlines between regions. The advantage of contouring is that this approach is insensitive to the absolute levels of gray (the "DC component") in an image and, consequently, to the absolute brightness in the illumination of a scene. That is, rather than looking for similarities in gray scale (as in the case of the homogeneous region approach) one looks for distinct differences in the gray levels of adjacent pixels.

In the contouring approach one distinguishes between <u>edges</u>, that is, borderlines between distinct homogeneous regions, and <u>lines</u>, that is, a contiguous train of pixels in an otherwise homogeneous region. Figure 4 shows an example of gray level image contouring and thresholding a gray level image, respectively.

The transformation of a gray scale image into a binary image is accompanied by a considerable information reduction, as illustrated in Figure 4.

Mathematically, one may view image segmentation as a <u>characteristic function</u> defined as follows. Let p(i,j) denote a pixel of an image and let C be the set of pixels forming a contour. The E(i,j) is a binary-valued function such that:

$$E(i,j) = \begin{cases} 1 \text{ if } p(i,j) \text{ belongs to } C \\ 0 \text{ if } p(i,j) \text{ does not belong to } C \end{cases}$$

<u>Segmentation into Edges</u>

There exist a variety of <u>edge detection operations</u> that perform exactly this kind of segmentation, the best known ones being:

(A)

(B)

(C)

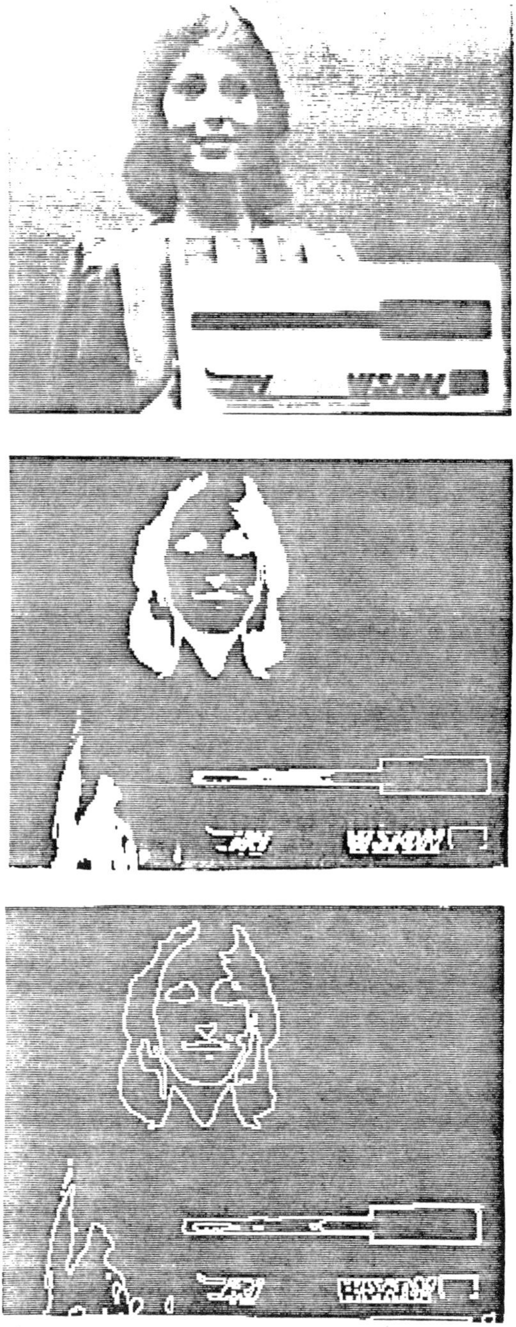

Figure 4. (A) A gray level image.
(B) Region image obtained from (A) by thresholding.
(C) Contour image obtained from (A) by Roberts Gradient.

- Roberts gradient operator
- Sobel operator
- Laplacian operators
- Prewitt operator
- Hueckel operators

The Laplacian operators, Prewitt operator, and Hueckel operators all are members of a larger class of operations called convolutions. Convolution is a generic operation that can be used to accomplish many different effects in image enhancement and segmentation (e.g., filtering, feature extraction). Convolution of an image by a 3x3, 5x5, or 7x7 matrix is a built-in function of the array processor in the IRI Vision Systems.

In contrast to convolution, which is defined as a linear function, the Roberts gradient and the Sobel operator are non-linear functions and, thus, do not belong to the class of convolution operations. Consequently, they are incorporated in the array processor as separate functions.

Segmentation into Regions

In the past, the most popular segmentation function has been the thresholding function defined as follows. Let x(i,j) denote a pixel value and let L be a given gray scale level. Then the thresholding function T(i,j) is defined as a binary-valued function such that:

$$T(i,j) = \begin{cases} 1 \text{ if } p(i,j) \geq L \\ 0 \text{ if } p(i,j) < L \end{cases}$$

The thresholding function is very popular because of its simplicity. In the past, when neither the large memory nor the high processing power needed for gray scale image processing was affordable, image analysis systems of the "first generation" were by nature binary systems. That is, image acquisition was directly combined with image thresholding and, consequently, image analysis started out from a binary representation of the image. Needless to say, much of the useful information contained in a gray level image may have been lost right from the beginning, and feature enhancement filters, which in many applications are a precondition for successful performance, could not be applied.

If an image can be acquired such that the features one is looking for contrast very decisively with the rest of the image, then it may be suitable to begin the image analysis procedure with a thresholding. But even then it is useful to have the gray level information about the image available in order to determine the optimum threshold selection. Moreover, gray level processing capabilities become absolutely mandatory whenever the significant features in an image cannot be extracted without the use of sophisticated feature enhancement functions.

The segmentation into regions by thresholding leads to spatial regions of adjacent pixels that are all one. Each such region is called a _blob_. A blob corresponds with a region in the original image consisting of pixels whose gray levels are all in the same gray scale interval.

Often the next step in the feature extraction procedure is to calculate certain parameters of a blob, such as its area, center coordinates (centroids), and second moments of inertia. In a scene consisting of many blobs, each blob must first be singled out before its parameters can be calculated. A most efficient approach to singling out such blobs is a procedure called _blob labeling_ and described later.

Feature Extraction

One powerful method of feature detection - template matching - has already been discussed briefly. In template matching, the image is correlated with a template of the desired feature, and a match is indicated by a maximum in the correlation function. If this method is to be accurate, one may have to use the normalized cross-correlation function, the mathematics of which we cannot discuss here without exceeding the scope of this introductory primer. The cross-correlation is quite expensive in terms of the number of arithmetic operations to be performed.

Therefore, only systems with very powerful array processors can handle the complexity of matched filtering in real-time.

Because of the computational complexity of template matching, one often resorts to simpler methods of extracting significant features from the gray level distribution over a pixel region or from a binary segment.

Examples of features extracted from the gray levels of the pixels of a region are:

- Histogram and statistical parameters of histograms
- Profiles and statistical parameters of profiles

The histogram is the gray level density distribution, i.e., a function that represents the frequency of occurrence of each gray level.

The profile is a spatial distribution of the average gray level of the pixels along a line. Statistical parameters are such moments of a distribution function as mean, variance, etc.

Examples of features extracted from (binary) image segments are:

- number of pixels of the segment;
- first moments (centroids) of the segment;
- second moments of the segment;
- distribution of the 0's and 1's in certain regions of the segmented image.

Especially in the case of blobs, one may calculate the number of pixels of the blob (corresponding with the 0th moment or area); the center coordinates (first moments); the circumference; and the orientation; the latter being derived from the second moments according to Steiner's law.

Another approach to extracting features from binary blobs is to apply <u>shrinking algorithms</u> for skeletonization or erosion.

In skeletonization, the blob shrinks toward its median line or "skeleton". If the median line is supposed to be a straight line, the next step may be to perform a least square fit and, subsequently, calculate the start point and end point of the straight line obtained.

In the procedure called <u>erosion</u>, a blob is shrunk even further, toward a cluster of pixels or even a single pixel. The result obtained may be indicative of the shape of the original blob. For example, if the blob is an ideal circle, application of an appropriate erosion operator may shrink it into the center point. If, on the other hand, the erosion operator yields a cluster of pixels rather than a single one, it is indicated that the blob is not a circle. Hence, erosion operators can be used to yield information about the <u>morphology</u>

of objects. Morphology analysis works particularly well with rotationally symmetric objects.

Blob labeling primarily is a means of separating different blobs by labeling them differently. If there exists only one single object in the first place, feature extraction can be performed in principle directly on the binary representation of the object, rather than going first through the elaborate procedures of blob labeling and connectivity analysis.

However, such binary representations usually are far from being without flaws, which may exist in the form of gaps (pixels that should be white but are black) or noise (pixels that should be black but are white). To eliminate gaps, a dilation (or "fusion") operator may be employed, which causes separate clusters of pixels to grow together, thus filling the gaps. Consequently, erosion often is combined with a preceding dilation.

Dilation and erosion are simple logical operations performed on binary image segments. Therefore, they are quite popular in simple image analysis systems that have little - if any - high-speed arithmetic capabilities. Often, the simplicity of this approach must be paid for with high noise sensitivity and, hence, low robustness. Noise in the image can readily cause the dilation operation to create artifacts, which the subsequent erosion will not eliminate. Statistical measurements, such as the moments of inertia, on the other hand, are averaging processes with "innate" noise immunity.

Image Understanding

At this point in the sequence of steps forming an image analysis procedure, a transition occurs. The previous steps resulted either in transformed pixels or, eventually, in features extracted (Figure 1). These steps of the image analysis procedure are referred to as the iconic phase of image analysis. From here on, the features extracted are called "symbols", and the ensuing steps of the image understanding procedure are referred to as symbolic processing.

From a preceding training phase the image analysis procedure may have been armed with a priori knowledge, given in the form of predetermined symbol values and ranges of symbol values, respectively. The acquisition of such a set of values and value ranges by training has the purpose of providing a model for a correct "scene" such as a correctly stuffed printed circuit board, a flawless object, etc. A typical training procedure may consist of showing the vision system a correct object or scene and pointing out all significant features.

During each subsequent recognition phase, the vision system extracts the same features and compares them against the ones represented by the model. In order to enable the system to decide whether a recognized object or scene and its model match or deviate in significant features, an acceptance/rejection criterion must have been established under test conditions.

Before proceeding further, let us put the problem in perspective. In a typical part inspection task, a manufactured part, e.g., a metal washer, might be inspected for acceptance on size and concentricity. An image of the washer is analyzed, and the features of area, inner and outer radius, and radius deviation are extracted.

The vision system must reach a decision of acceptance or rejection, based upon the measurements as well as on a priori knowledge. While each feature is compared individually, the overall decision might involve a complicated comparison routine. For each individual part a measurement is compared to an *a priori* defined threshold value, below (or above) which the feature is rejected (or accepted). Note that if the outcome of this comparison is acceptance, it may mean that the part in reality is acceptable or that a very similar but unacceptable part has passed. Analogously, if the outcome is rejection, it may mean that the part is in reality unacceptable or that a very similar but acceptable part has failed. The lower the acceptance threshold is set, the more likely it becomes that rejectable parts in reality are accepted. Conversely, the higher the acceptance threshold is set, the more likely it becomes that acceptable parts in reality are rejected.

Thus the selection of appropriate acceptance thresholds is one of the crucial decisions the designer of the test has to make. The threshold values used in acceptance/rejection decisions should be based upon a representative range of examples and rejection rates. Total acceptance will depend on the analysis of tests performed on all the features. Image understanding in its simplest form consists in the performance of individual feature tests based upon thresholds that are determined *a priori*.

As our discussion may have shown, image understanding can be a fairly "intelligent" procedure that begins with training and the construction of a model, in a specific data base of the system called the "model base", and ends with the recognition of objects, features in objects, or complex scenes by comparing extracted features with the model and reaching a decision.

Systems that acquire knowledge from human experts and ensuingly use it as a basis for cognitive faculties such as recognition, inferencing, and decision making usually are called "expert systems". The image understanding system described above may be viewed as a (simple) *model-based expert*

<u>system.</u>

A robust decision process requires that "uncertainty regions" are defined in the feature space in which feature recognition and thus classification is not possible. There exists a dichotomy between the size of the uncertainty region and the correctness of decisions (classifications): the larger this "gray zone", in which the system has to confess that it cannot reach a decision, is allowed to be, the smaller is the probability for a wrong decision.

On the other hand, if in too many cases the answer of the system is neither "yes" nor "no" but "don't know", this may become a nuisance for the user. The frequency of such "don't know" cases often can be reduced by adding rules that enable the system to reach the right decision even in the case where the underlying recognition process has no definite result. Adding such rules leads to what is known as a <u>rule-based expert system.</u> The very complex image understanding tasks that will be encountered in the future in connection with computer integrated manufacturing (CIM) probably can hardly be carried out without the aid of rule-based expert systems. However, as was mentioned before, this area, as yet, is very much in its infancy.

In important cases the task may be to find the <u>location</u> of an object represented by a blob. Here, the calculation of the first and second moments of inertia is most helpful, yielding center coordinates and orientation. In other tasks, the location of an object may be <u>a priori known</u>, but the type of object actually found in that location must be recognized. Examples of such applications are: optical character recognition, printed circuit board inspection, etc. In this case, one may calculate some of the other parameters mentioned above over a region given by a "window of expectance", whereby the window is set according to some a priori knowledge about the expected location. The parameters obtained then will be used as input to a classifier procedure that performs the actual recognition.

CONCLUSION

Now that you are well on your way to locating football players from television images and rejecting faulty washers based upon a priori thresholds, let us review for a moment the elements of this chapter.

Whatever the task at hand, here are some "golden rules" as to how to proceed after you have defined your image analysis task:

1. Get a good image first. Why make the "little man" inside the Vision System box work harder than he should?

2. Play with as many enhancement procedures as you can conceive before feature selection, using your own human vision system as guide. But don't let yourself be fooled by enhanced images without enhanced features!

3. Examine a wide range of image examples before selecting an acceptance/rejection threshold. Feel confident of the "robustness" of your algorithm, but don't hesitate to revise it if it fails on significant cases.

4. After you have developed an algorithm, examine it for proper utilization of the hardware speed of the vision system (especially if your system contains a co-processor). List the execution times of the instructions in your program. Do they add up to the shortest possible overall time? Could their order be modified?

Tactile Sensing for Robots

Jack Rebman
Lord Corporation

In addition to visual sensing, tactile sensing has recently been developed to the point where commercial devices designed for robot feedback and control have become available. This chapter explains the fundamental principles behind tactile sensing, the several classes of sensors that are available, and potential applications to which this new technology may be addressed.

Mr. Jack Rebman received BS and MS degrees in mechanical engineering from Case Western Reserve University and UCLA, respectively. He is a Senior Fellow, Technology at the Lord Corporation Research Center and a visiting lecturer at Duke University. His research programs have emphasized the development of systems to provide dextrous and intelligent manipulation, with emphasis on sensors and end-effectors and he has made frequent technical presentations at technical meetings in the field of tactile sensing.

Tactile Sensing for Robots

A Context for Tactile Sensing

Automation has been a practical reality for many years and its application has resulted in significant benefits. Because of the specific capabilities in available automation technology those benefits have been limited to a restricted set of manufacturing tasks and processes. High levels of consistency and precision in work pieces and high levels of repeatability and accuracy in the manufacturing equipment have been required. Even though consistency and precision might not be needed for product function, they are musts for most automated manufacturing methods. This has contributed to increased costs of automation. Economic justification can be shown only for large quantities of production. Therefore, products manufactured in batches of different work piece size, model, configuration, etc., have not generally benefitted from automation. This is especially true for discrete products that do not normally maintain constant shape, such as limp textile materials.

Progress has been made on the problem of automatic manufacture of batches of different characteristics by systems that are programmable. These are machine systems that can be conveniently directed to change their mode and sequence of operations. Normally by means of software instruction, they can be made to perform the same operations on a different work piece or perform different operations on the same work piece. Numerical control machine tools and currently available industrial robot arms are excellent examples.

These programmable systems still require high levels of consistency, precision, and repeatability. In addition, they need one other performance capability-adaptability. An adaptable machine is one which can alter its own operations to accommodate a range of variations. These variations can occur from one work piece to another as well as in machines, fixturing, and other factors in the work environment. Io be adaptable, the machine must be able to get information about the operational situation so as to select a proper next step from available alternatives.

This means that the machine must (1) have the ability to collect data about important variables in its environment, using suitable sensors, (2) develop conclusions about the operational meaning of the newly sensed information, (3) make a selection about the next action to be taken, (4) generate suitable control commands, and then (5) perform the action. This sequence would be repeated until the desired task is completed.

What is evolving here is a description of an adaptive manipulation system having some level of artificial intelligence. The diagram in Figure 1 represents a general adaptive manipulation system. The four items in the center are sub-

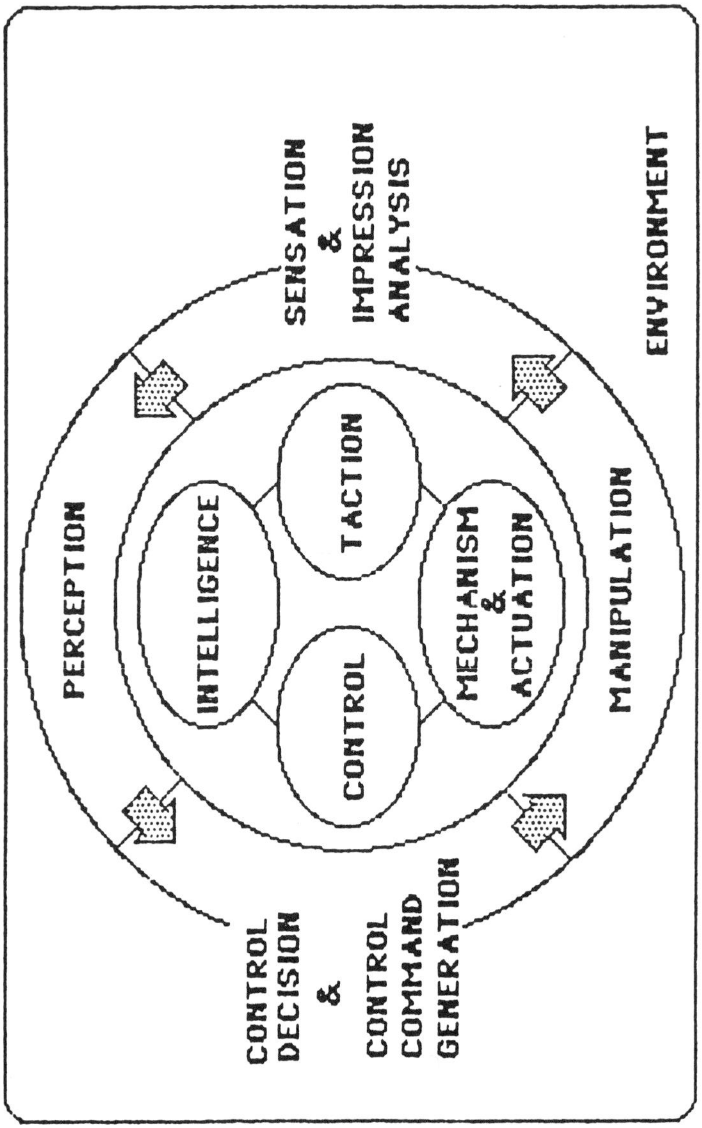

Figure 1. General adaptive manipulation system.

systems which perform the functions labeled in the outer ring. The arrows indicate the sequence in which the functions are generally performed.

The actuation and mechanism comprise that portion of the system which performs the mechanical actions of the manipulation. It could take a variety of forms from a simple two-fingered gripper to a more nearly anthropomorphic hand with fingers and thumb. The taction is that portion of the system which collects and processes information about the internal mechanical and external contact conditions. Here too, taction can be embodied in a variety of forms to gather information from the contact environment and extract significant features from it. Depending on the performance required from the system, the taction subsystem could consist of simple sensors for detecting presence to complex arrangements with multiple sensing modes.

The intelligence is that portion of the system which integrates all of the taction information and draws conclusions about the condition of the environment, the status of the system, and the relation of these to the manipulation objectives. Based on conclusions regarding the system and the environment, it would select a course of action for the manipulation system. This intelligence capability could be vested in hardware or software, either of which could be general or dedicated. The control subsystem then develops and sends the commands for appropriate continuing actuation. The forms of control could vary from simple on-off controls to complex servoed systems with hierarchical architectures.

Only when all of these major subsystems are present are the conditions met for having an adaptive manipulation system. The large majority of the robot arms in industrial use are not adaptive. In fact extremely few have the necessary capabilities. They do not possess suitable sensing capabilities nor the elements of artificial intelligence which enable the machine to draw a perception of the status of its work environment and make a decision about what to do next.

The emphasis in this chapter will be on the sensing and what might be done with the data collected. In particular, the focus will be on contact sensing, that is tactile sensing or taction as some people prefer to call it.

Tactile Sensing

The definitions of tactile, touch, and taction suggest that the three are essentially synonyms. Common usage seems to support this. Common usage also assigns many other meanings to the word touch. They range from a light stroke, tap, or push to the act of inducing to give or lend money. The following meanings are taken from Webster's Ninth New Collegiate dictionary (Merriam-Webster).

Tactile--1: perceptible by touch 2: of or relating to the sense of touch.

Touch--1: to bring a bodily part in contact with esp. so as to perceive through the tactile sense; handle or feel gently usually with the intent to understand or appreciate.

Touch--3: the special sense by which pressure or taction exerted on the skin or mucous membrane is perceived 5: a specified sensation that arises in response to stimulation of nerves sensitive to touch.

Taction-- touch.

These definitions are pertinent to the subject of this chapter.

Touch information is obtained through physical contact of

the sensor with a target object. Basically the sensor returns signals as a function of the forces and the deflections resulting from the contact impression. In turn, these can be interpreted as characteristics of the contact interface, as features of the contacting object, and as conditions of contact. Examples are:

1. contact or presence
2. target shape, location, orientation
3. contact area, pressure, pressure distribution
4. force magnitude, location, direction
5. moment magnitude, plane, direction

This list can be expanded by including mass and inertia effects, dynamic as well as static conditions, and combinations which define for the target, its compliance, texture, etc. Any particular sensing system might deliver one or various combinations of the segments of the above information.

Contact information can be divided into two general domains of interest, (a) geometry and (b) forces and moments. Presence, shape, location, and orientation fall within the geometry domain. Pressures, forces and moments fall within the force-moment domain. Tactile sensing, as a contact sense, is uniquely suited to obtain information from both domains. Perhaps, this can be illustrated by making a comparison between machine taction and machine vision.

Machine vision is a non-contact sense. It collects information from a stand-off position. Whether the stand-off distance is large or small, it always collects information in an indirect mode. Normally, it obtains global information with the potential for including myriad details. It gets the big picture, as a two-dimensional representation of a three-dimensional world. With the absence of physical contact, it cannot sense forces and moments, except, perhaps, with some exotic techniques. Machine taction, on the other hand is a contact sense. It collects information while in direct physical engagement with the target object. The field size of its contact impression is limited and is best suited for conveying local information. Taction would have great difficulty in getting the big picture. It can readily collect dimensional information. Also, since it does physically make contact to perform its function, taction can measure forces and moments. Taction, therefore, can read from both the geometry and the force-moment domains.

This brief comparison begins to show what a more detailed look would indicate. Machine vision and machine taction are largely complementary senses. A major portion of sensing functions are those which one sense, taction or vision, can perform considerably better than the other sense. Many industrial tasks, such as assembly, require both of them, used in a coordinated way. Certainly, in some fewer situations, the sensing functions can be performed by either vision or taction, in which case they are competitive.

Tactile Sensor System

A tactile sensor which reads patterns of information resulting from loads on the touch surface is shown in Figure 2. The sensing unit is in the foreground and a display of its output is in the background. This particlar model is sensitive only to loads perpendicular to the touch surface. The touch surface includes an array of sensitive sites. Each of these sites outputs a signal proportional to the deflection at the site. The display on the video monitor is a representation of the touch surface, indicating the 8x8 array of sensitive sites. In the display, the vertical line extending upward from a given site location is an indication of the amount of deflection at that site.

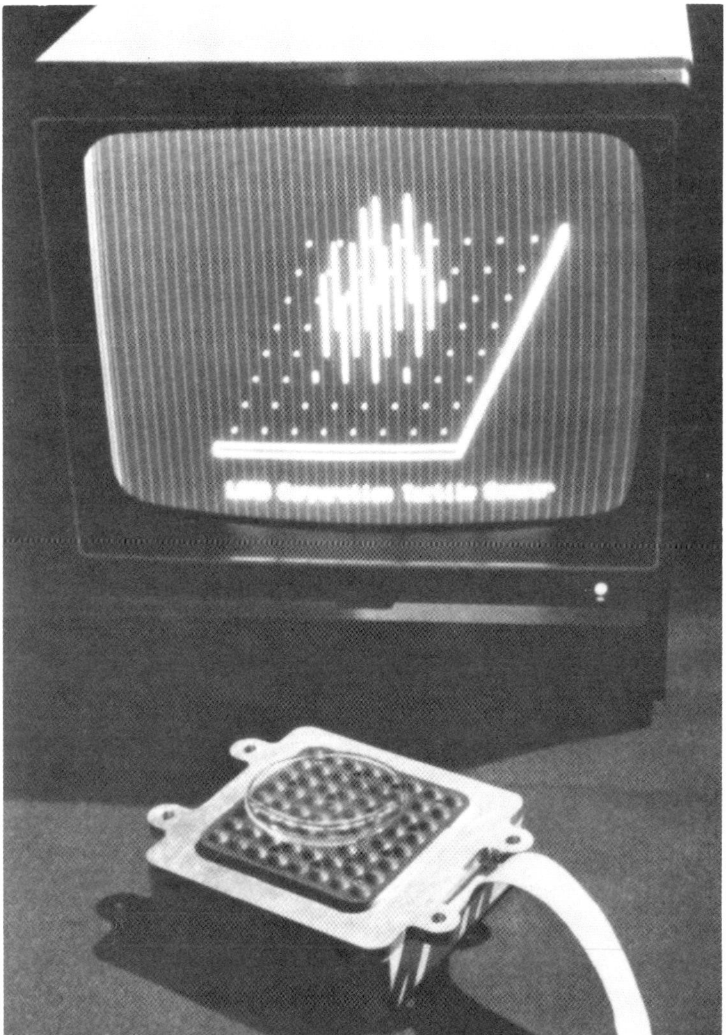

Figure 2. A tactile sensor with display of tactile impression.

The target object in the photograph in Figure 2 is a piece of transparent material having a spherical shape in contact with the touch surface. The impression made by such a surface in bearing against the sensitive sites is shown as displayed on the monitor. To obtain the magnitude of deflection shown in the display, this particular sensor would require that an external force be applied to the target object, pressing it against the touch surface. The information in this tactile impression can be used to describe many aspects of the contact situation. The significance of that information will be discussed later.

The sensor shown in Figure 2 includes all of the major components of a tactile sensor system. These are:

1. touch surface
2. transduction medium
3. structure
4. control/interface

These items also identify major functions within the sensor.

Touch Surface

Figure 3 is a line drawing of the sensor unit. The sensitive sites are shown as projections of the touch surface. This is only one of many useful configurations of the touch surface.

The touch surface is the sensor's window to its world of interest. Through it, it obtains information pertaining to the geometry of that world and also information pertaining to the load situation prevailing. Besides being a window, the touch surface acts as a shield from harmful agents.

Information pertaining to geometry contributes to drawing conclusions about features such as location, shape and alignment of the target object. To obtain sufficient data regarding

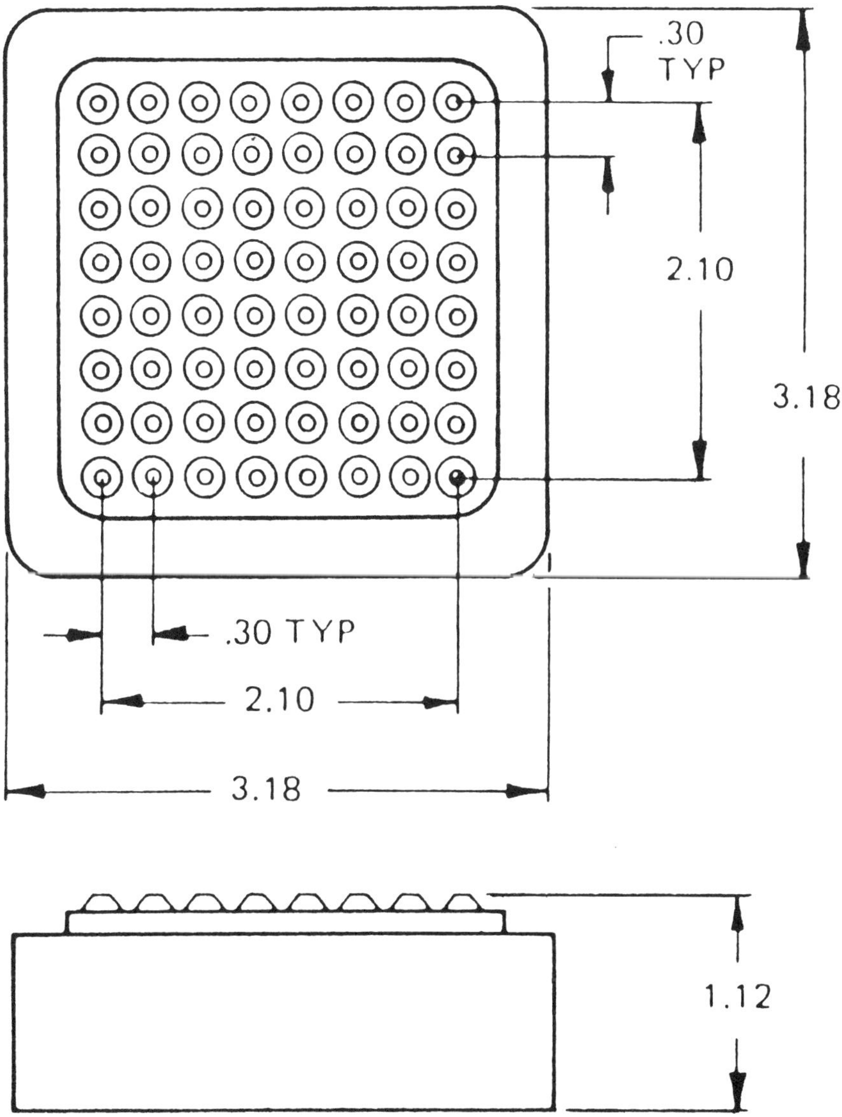

Figure 3. Tactile sensor unit.

these features, the touch surface is required to conform to appropriate portions of the object. Low compliance, a hard touch surface, can limit the amount of contact information that could be obtained. This is true for any shape that does not perfectly match the touch surface contour or is misaligned relative to that surface. The compliant layer enables getting contour and shape information both in the plane of and also normal to the plane of the touch surface.

As a compliant layer, the touch surface behaves as a spring. Objects bearing against a touch surface produce deflections in relation to the contours and location of that portion of the target object in contact with the sensor. At any particular point of interest within this contact area, the local load or pressure will be the product of deflection and the spring rate of the touch surface at the point of interest. The spring rate of the compliant surface is an important design parameter affecting performance of the overall sensor system.

The stability of grasping is an additional factor on which the touch surface has an important influence. The compliant touch layer enables development of much higher shear loads on the touch surface than would be available based on the coefficient of friction alone. The "digging in" provides areas of direct bearing through which shear loads can be carried. Shear loads are those parallel to the touch surface. This is important for gripping with an end-effector mounted sensor, even though shear loads might not be measured by the sensor.

Several avenues are available for obtaining desired compliance. One is through material selection. Choices among basic elastomers, for example, and particular formulations of each can provide a wide spectrum of spring rate characteristics. Coupled with a particular material, significant control of compliance can be achieved through configuration design of the layer comprising the touch surface. The tactile sensor in Figure 2 makes use of both avenues, material selection and design.

Transduction Medium

The basic role of the transduction medium is the conversion of the mechanical information at the touch surface into electrical signals which can be handled and processed conveniently. The transduction can take a variety of forms, many of which have been reported in the literature.(1) These include piezo-resistive, piezo-electric, conductive, magnetic, capacitive, etc., materials and devices. Virtually any approach which exhibits a transduction effect between mechanical strain and another form of energy can make a useful tactile sensor. The utility of one transduction medium over another depends on many aspects of performance and economy requirements.

For the tactile sensor illustrated earlier, the transduction from mechanical to electrical signals is optoelectronic using solid-state devices. A section view of a typical transducer site is shown in Figure 4. At each sensitive site within the array, an emitter and detector look directly at each other across a short distance. The light beam across the gap can be intercepted and shaded partially or fully by a projection from the touch surface layer. Deflections on the touch surface cause more or less shading and in turn, more or less electrical response. This concept of transduction for tactile sensing was first reported in the literature(2) in 1973.

The solid state devices used are an LED emitter and a photo transistor detector. The "C" dimension indicates the lens aperture on each of the optical devices. As an object

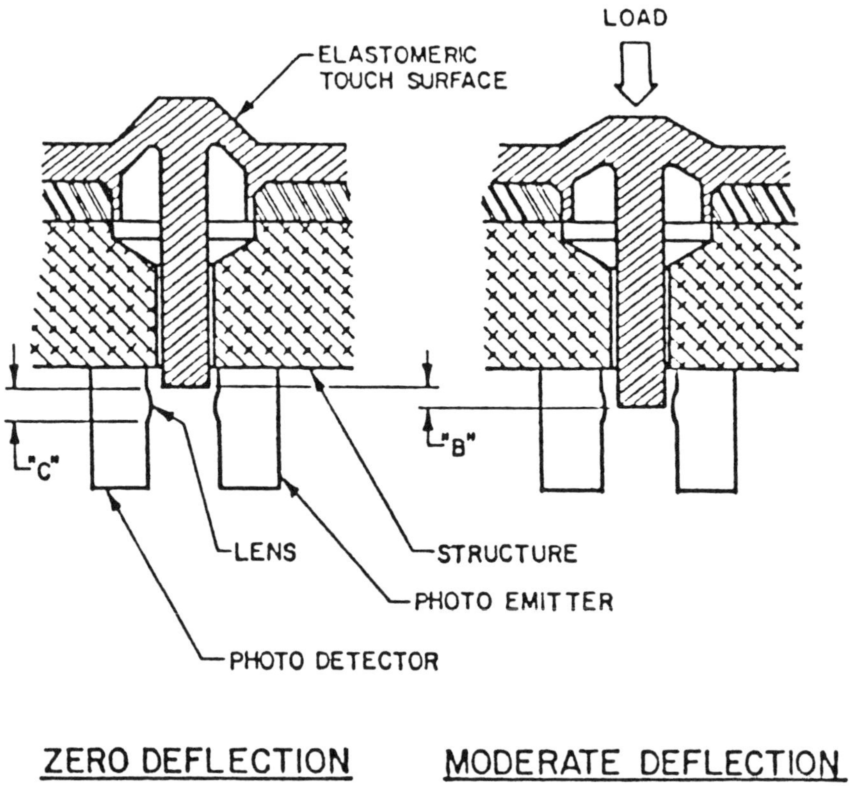

Figure 4. Section views of a typical transducer site with optical electronic transduction, showing no load and load conditions.

contacts the compliant touch surface above this emitter/detector pair, the surface deflects as a function of an externally applied force or an imposed deflection. The two conditions, undeflected and deflected, are illustrated in Figure 4. Dimension "B" defines the magnitude of the deflection. The shading element, integral with and projecting inward from the surface, is thereby moved between the emitter and detector changing the amount of light passed by the photo transistor as a function of the movement of the shading element. The value of the current can be read as the deflection of the touch surface.

It should be noted that in reality, the transduction medium measures only deflection. Force is inferred through use of the force-deflection characteristics of the compliant touch surface.

Transduction uniformity is very important for taction based adaptable manipulation. Each sensitive site is a separate force/deflection sensor. Ideally, the performance of each would match every one of the others. Practically, of course, this is not the case. Mechanical, electrical, and optical factors introduce variations in performance from one site to another. Examples include dimensional tolerances of the individual parts and assemblies, reflectivity in the locale surrounding the emitter-detector pair and electrical characteristics of individual emitters and detectors. Corrections for these variations can be made through use of appropriate hardware and also software. Both approaches are employed in the sensor in Figure 2.

The units of the emitters and detectors, such as those in Figure 4, are not identical. An inspection process is employed to match the members of each pair. A further step involves circuit tailoring at each site with the installation of an appropriately sized resistor so that the voltage output from any given site will be more nearly like that of the others. Once the hardware corrections are in place, the major adjustments are accomplished through software.

The curves in Figure 5 are based on measured data from the 64 sites of the sensor in Figure 2. The band of response variation for the uncorrected sites is shown first. The mean of the population within the band is the standard for correction. The result of both hardware and software corrections is then indicated, as in Figure 6.

Structure

The structure is that framework which ties together the various mechanical elements of the sensor device. Besides keeping the elements in a stable relationship, one with the others, the structure serves to package the elements and defines several other important features. Among them are: envelope dimensions, shape, weight, mechanical attachment, routing avenues for wiring, protection from the environment, etc.

The primary purpose of the structure is to support the various elements of the sensor device. It is important that satisfactory load paths be maintained between the touch surface and the transduction medium and among transduction sections for sensors with multiple modes of capability. Strength along these load paths must be adequate for application requirements, especially potential overload transients. Structural stiffness is as important as the strength. Stiffness contributes both to load path control and to proper dimensional stability among the sensor elements.

Depending on the style of the sensor device, the weight can be very important. It is desirable in many cases to install the tactile sensors as part of the end-of-arm tooling on

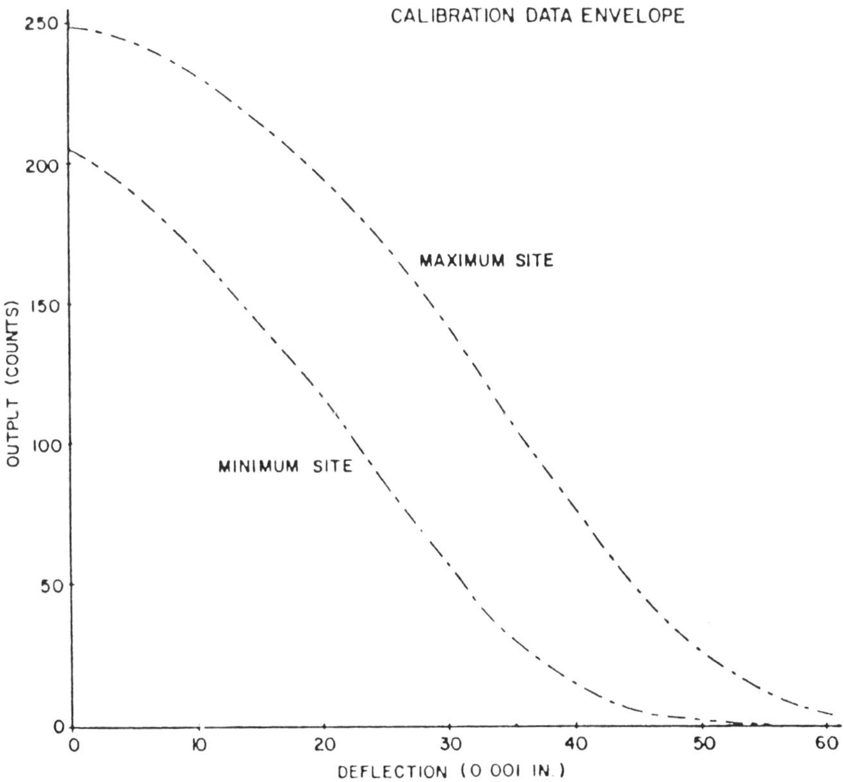

Figure 5. Calibration data envelope for 64 sites of an array sensor.

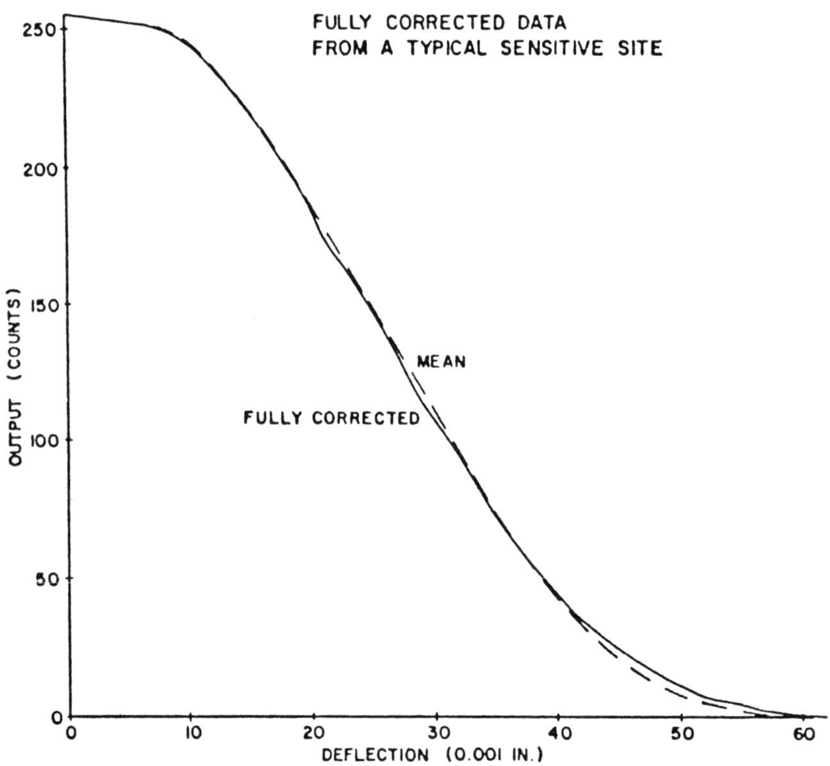

Figure 6. Fully corrected data from a typical sensitive site compared to mean of all the sites.

a robot arm. Just as with any machine that must carry a payload, the robot arm has a limited amount of weight that it can carry on the end of the arm. The sensor, as part of such tooling, is tare weight and uses some of the payload capacity. The weight as mass also affects the speed of operations. Acceleration and deceleration performance of the robot arm are directly related to the total payload mass.

Shape and envelope dimensions can render an otherwise well performing tactile sensor as useless. This is especially true for sensors which are mounted on end-of-arm tooling at the areas in direct contact with the workpiece. They can get in the way, preventing tooling access in confined working spaces. The means for attaching the sensor to whatever is the host equipment can have a significant effect on the sensor envelope and shape. Structural integrity, convenient access, and easy installation and removal are important factors for the attachment.

In order to perform its function, a tactile sensor must be immersed in the work environment. It must be physically placed in contact with the workpiece repeatedly over the surface of the sensor. This exposes the sensor to various contaminants, debris, heat, and electromagnetic radiation which can seriously degrade its performance. Although the touch surface bears, perhaps, a heavier burden in this regard, the structure has a large influence on the vulnerability of the sensor device to these environmental factors.

Interface

The fundamental purpose of the interface is to provide for the communications between the sensor device and the host equipment. In addition, it can provide a number of other functions that could reside in either the interface or in the host equipment. The other functions include such things as:

control and interrogation of the sensor device, signal conditioning, temperature compensation, calibration corrections, data processing, algorithmic computation, machine intelligence, and other capabilities not provided by the cooperating robot arm system.

Tactile sensors generally have a number of transduction elements in one or a number of forms. Raw data must be collected from these elements in a properly organized way. The sequencing and timing are functions of the interface. Most transduction media produce an analog output. The interface converts the signals from analog to digital and sets voltages and format to be compatible with the host system.

Many transduction media are temperature sensitive. Suitable compensation must be provided, many times by the interface, in order to make the output independent of ambient temperature. Calibration corrections adjust the output from each of the transduction elements. Differences in performance among those used is an expected reality. The problem is enlarged when, as in some tactile sensors, there may be hundreds of transduction elements. Two correction approaches can be taken. One is to adjust circuit components. Another is to make software corrections based on calibration data, as explained earlier.

A typical sensor system architecture is shown in Figure 7, including the interface.

Tactile Sensor Characteristics

Mode pertains to the direction of sensor response relative to the touch surface. The basic modes are normal and shear. Normal is in a direction perpendicular to the touch surface while shear is in the direction parallel to the touch surface.

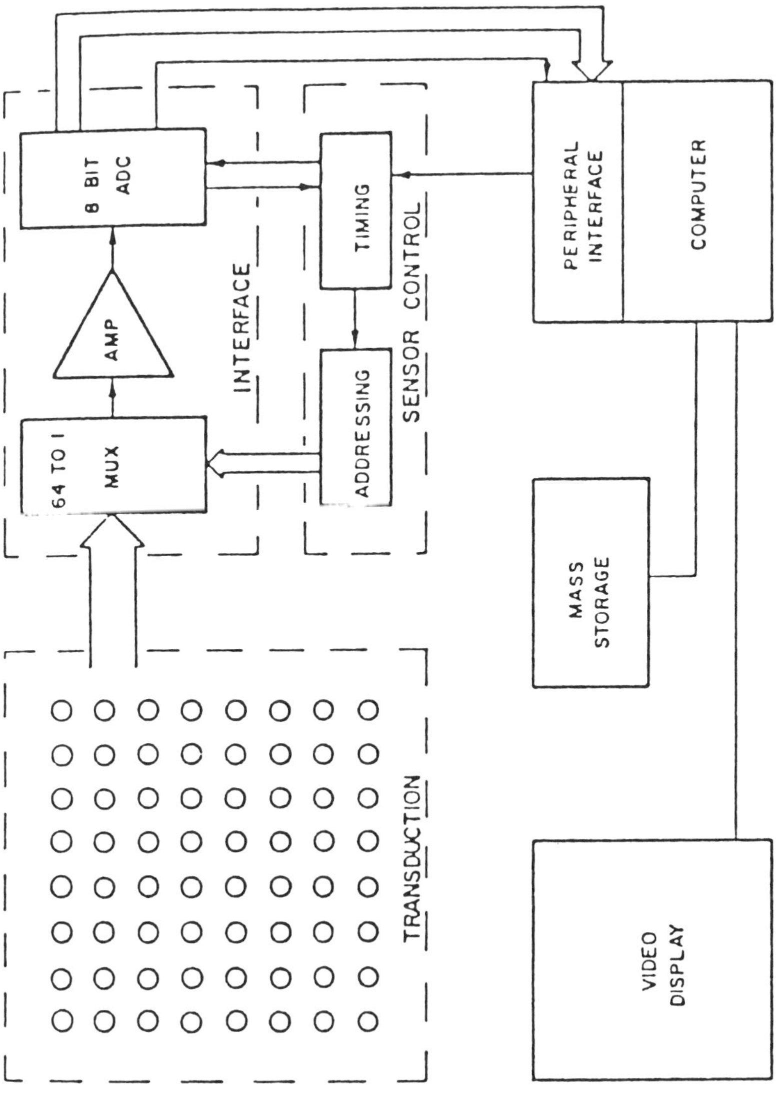

Figure 7. A simple sensor system architecture.

Depending upon the nature and distribution of sensitivity on the touch surface, determinations can be made of both forces and moments.

Mode can also indicate whether the sensor response is gross or is detailed so as to obtain readings of pattern. To provide pattern capability, multiples of sensitive sites are needed. The arrangement of the sites is generally a square or rectangular array, with individual site sensitivity oriented normal to the touch surface.

A further meaning of mode establishes whether the response is binary or gray scale. Binary is either on or off and merely indicates that the load or deflection is above an arbitrary threshold. Gray scale on the other hand gives a graduated output as a function of load or deflection and significantly increases the information content obtained from the contact impression.

The sensor of Figure 2 has, as noted earlier, an 8x8 square array of 64 sensitive sites for pattern information. Each site is sensitive only to normal loads and each has gray-scale sensitivity.

Some sensors measure the combined effect of all forces and torques imposed on the touch surface. Although this represents a gross measurement of conditions on the touch surface, many times the information is delivered in terms of force and moment components. These components relate to the six degrees of freedom for a particular coordinate system. Such a sensor can be called a force-torque sensor or a vector sensor.

A force-torque or vector sensor can be rendered in a number of configurations, depending on the size and the location of its placement in the manipulation system. That which generally is affixed to the end effector is a vector sensor. That which is inserted in the wrist of the manipulator arm is a force-torque sensor. The vector sensor is generally smaller than the force-torque sensor.

<u>Resolution</u> relates to several aspects of the sensor. First is the spatial resolution, which depends upon relative spacing of the active sites in a multi-site array. The spatial resolution is important for information about some aspects of geometry and location. A second type of resolution is at the individual site and indicates the gray scale resolving power over the range of load or deflection available at each site. This is important for describing the loading situation as well as additional aspects of geometry. This resolution is governed by several of the major components. For force inputs, the touch surface is the primary component. For deflection inputs, the transduction medium, and control and interface are the governing components. The most significant factor is the number of binary counts which are useful in the analog-to-digital conversion which might be employed. A third resolution consideration is in the time domain. Appropriately rapid system response to new conditions and to time dependent parameters is important.

<u>Sensitivity</u> is closely related to resolution and like resolution, it is governed by the analog-to-digital conversion. The sensitivity, before A/D conversion, is generally much better than after conversion. If so, once conversion is performed, the sensitivity equals the resolution. Sixteen-bit conversion rather than eight-bit would permit greater sensitivity.

<u>Range</u> is pertinent for the gray scale response at an individual sensitive site. The total range depends upon the dimensions and configuration of the touch surface as well as

the performance of the control and interface electronics. Generally, considerable overload capacity is available. However, under overload conditions, sensitivity is generally greatly reduced.

Contrast relates to the sharpness with which an edge can be imaged. In the taction sensor, it depends on the degree of load transfer from a given point of contact with the touch surface to adjacent areas of the touch surface. A number of factors contribute to the level of contrast. They include the touch surface configurations as well as the configuration of the transduction medium. In the sensor in Figure 2, a high degree of isolation of one site from another was achieved. This is true both mechanically and electrically. In fact, a potential disadvantage is inherent in the design as a result of this attempt to enhance contrast. Areas of insensitivity are present between sites which might miss information for sharp cornered targets contacting in that area between the sites. A modest amount of mechanical coupling from site to site would relieve this particular problem. Such a loss of contrast would have to be evaluated against application requirements.

Tactile Sensor Styles

Tactile sensor styles are determined by the manner in which the sensor is used and also by the size of the sensitive fields. There are several general classes of style: gripper mounted, work surface mounted, large field, and small field. The large and small field designations are two subclasses that would fit either within the gripper or work surface mounted styles. Large field sensors are likely to be work surface mounted.

Gripper mounted units are normally attached to the end effector of the manipulator system. As such, they are taken to the work place and collect data as the work piece is grasped by the end effector. Data collection can be continued so long as the end effector has hold of the work piece. For the work surface-mounted sensor, the device is normally attached to some stationary surface such as bench top or fixture. For this arrangement, the work piece is manually carried to and from the sensor. Once the work piece is placed on the sensor surface, data can be collected. For a work surface mounting, the sensor must generally be sufficiently sensitive so that the weight of the work piece alone will be sufficient to activate the transduction medium in the sensor system. This sensitivity requirement is more severe than that for the gripper-mounted device, since the clamping force of the gripper upon grasping the work piece can help make a tactile impression.

Figure 8 shows several models of a tactile sensor device. The left hand and central units are modules which can be used independently or as an integrated assembly. The integrated assembly is shown as the unit on the right. The module on the left is an array sensor. It includes a rectangular arrangement of sites in a 10 x 16 pattern. The device at the center is a vector sensor. This unit delivers information on the three components of force and of moment about the coordinate axes referenced to the touch surface of the sensor. The combined unit, in conjunction with the interface, can deliver the full set of array and vector information. Figure 9 shows a two-fingered gripper with complete array and vector sensor devices mounted on each of the fingers.

A work surface-mounted sensor is illustrated in Figure 10. The sensor shown is a square shaped field with 6400 sites in an 80 x 80 array. The work piece is a metal stamping which has been placed on the sensor touch surface by the gripper mounted on the robot arm. This sensor is also designated as a large field sensor. The spacial resolution of the sites is 0.080 in.

Figure 8. From left to right, array only sensor device, vector only sensor device, and combined array and vector sensor device.

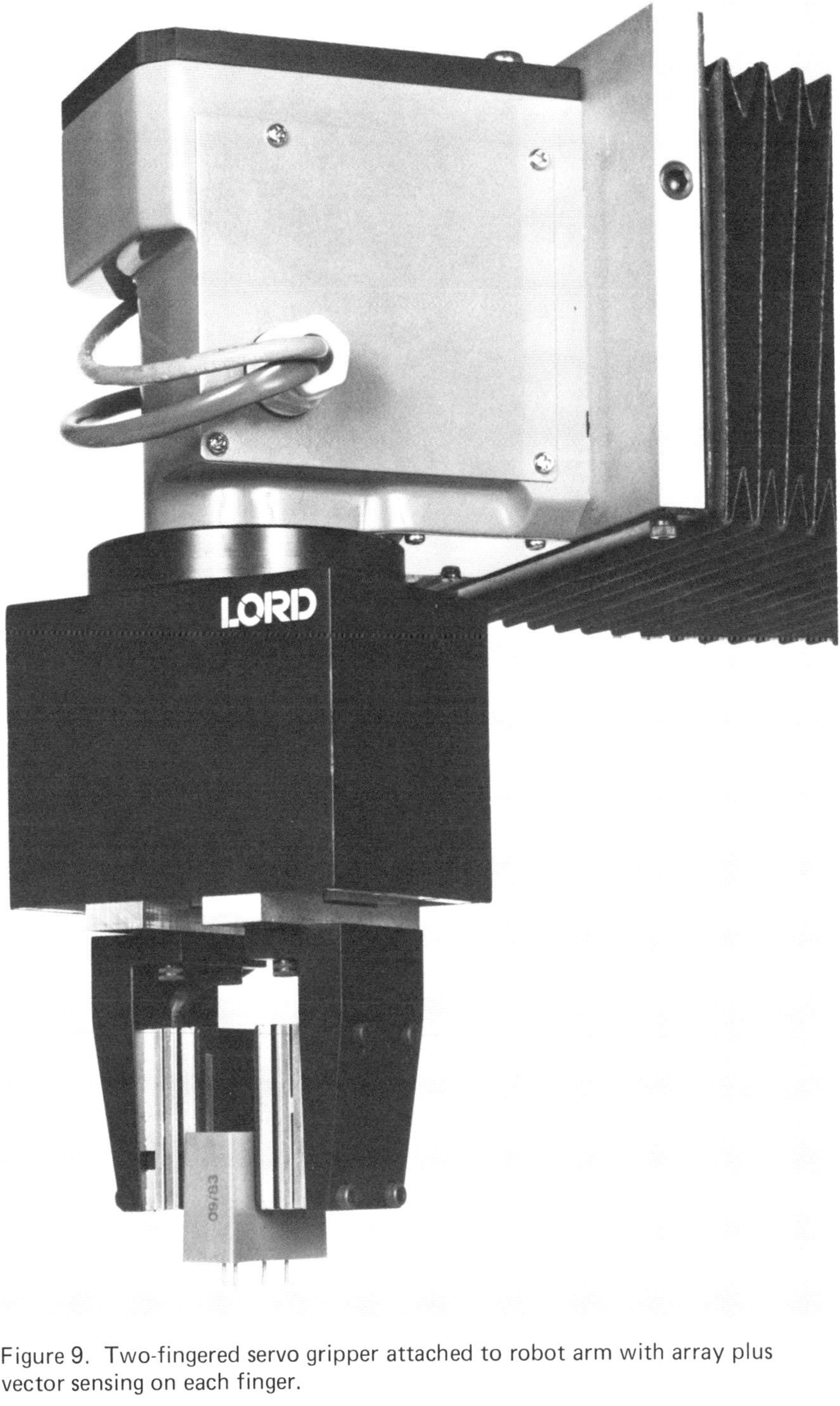

Figure 9. Two-fingered servo gripper attached to robot arm with array plus vector sensing on each finger.

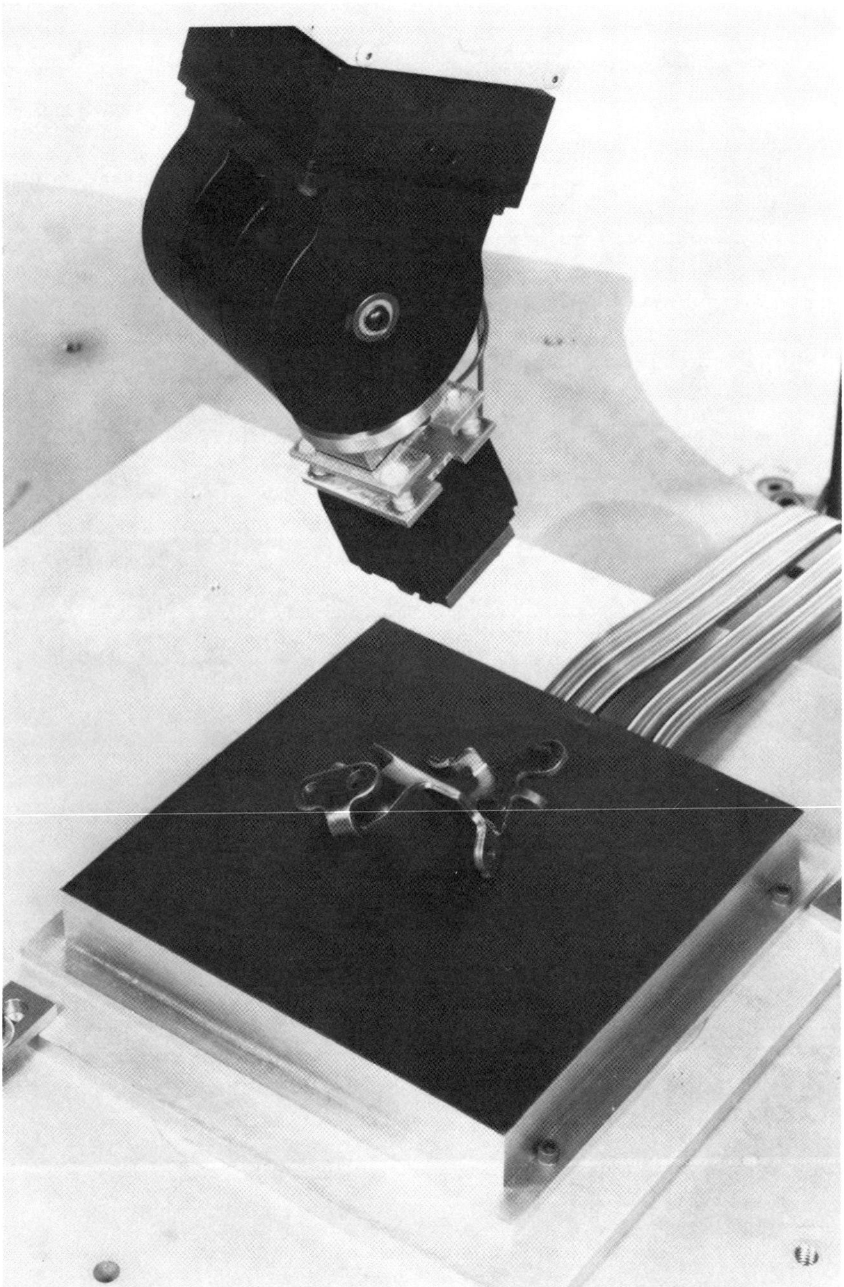

Figure 10. A large field work surface mounted tactile sensor.

The square sensitive field is approximately 6 in. on a side.

If the site spacing were reduced to 0.020 inches, the entire unit could be scaled down to the point where it might be conveniently affixed to the fingers of a gripper. Although it would be gripper-mounted, it should still be considered a large field sensor because of the number of active sites.

There is no clear differentiation between large field and small field in terms of the number of sites. Generally, 16 active sights per side, 256 total, of an array is a useful threshold between the two. The protocol for control of the sensor could further adjust that threshold one way or the other.

Sensor Application

In considering the utility of the data obtained with the taction sensor, please refer to Figure 2. Each of the dots in the presentation represent an individual site on the touch surface. The vertical lines extending from some of the sites represent the magnitude of deflection occurring at the particular site.

Several conclusions can be drawn from this pattern of lines and line lengths, about the target object and its location and orientation. The particular sites which are activated provide data which can be used to locate the target object and its location and orientation, and also to locate the target relative to a reference on the touch surface. The pattern of sites activated will define the size and shape of the contact area. The relationship of the upper ends of the lines extending from the activated sites can be used to describe the shape of the surface of the target within the contact area.

If the target object were flat over the contacting area, the ends of the lines extending from the activated sites would fall on a plane. Even though the ends of the lines fall on a plane, the lengths of the lines might change from long to short from one side of the contact area to the other. Knowing the target contacting surface to be flat, this would indicate that the alignment of that surface relative to the touch surface is not parallel. Appropriate computations based on the lengths of the lines would provide information about the actual orientation.

To this point, we have considered the line lengths in the display to represent deflections. By introducing the load-deflection characteristics of the individual sites, each line can also be interpreted as the load on each of the individual sites. This, then could represent a pressure pattern over the area of contact. The summation of all the line lengths of the activated sites would represent the total load on the touch surface. The centroid of that pattern would be useful in determining load and moment conditions imposed on the target object.

Taction sensors have been used in conjunction with a robot arm to perform assembly of a flexible diaphragm and plastic cap. The diaphragm and cap are sketched in Fig. 11 and Fig. 12. The diaphragm is made from an elastomer and is quite flexible. Its top and bottom surfaces are different which means there is a proper head and tail placement of the diaphragm in the cap. The cap is made from nylon. Because of the construction of the passages leading into and out of the cap, it has a right and left orientation. When assembled, the diaphragm must not only have the correct side up, it must be securely seated in a reentrant lip in the cap. After assembly, the unit must be placed in a nest, oriented for a subsequent operation.

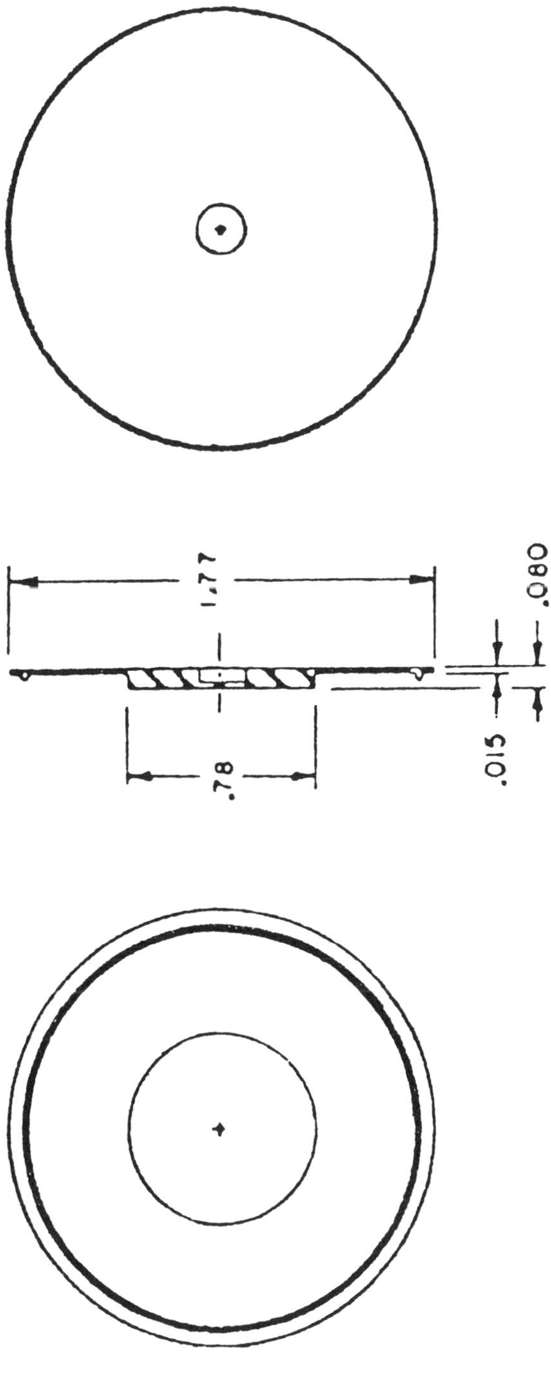

Figure 11. Flexible diaphragm assembly component.

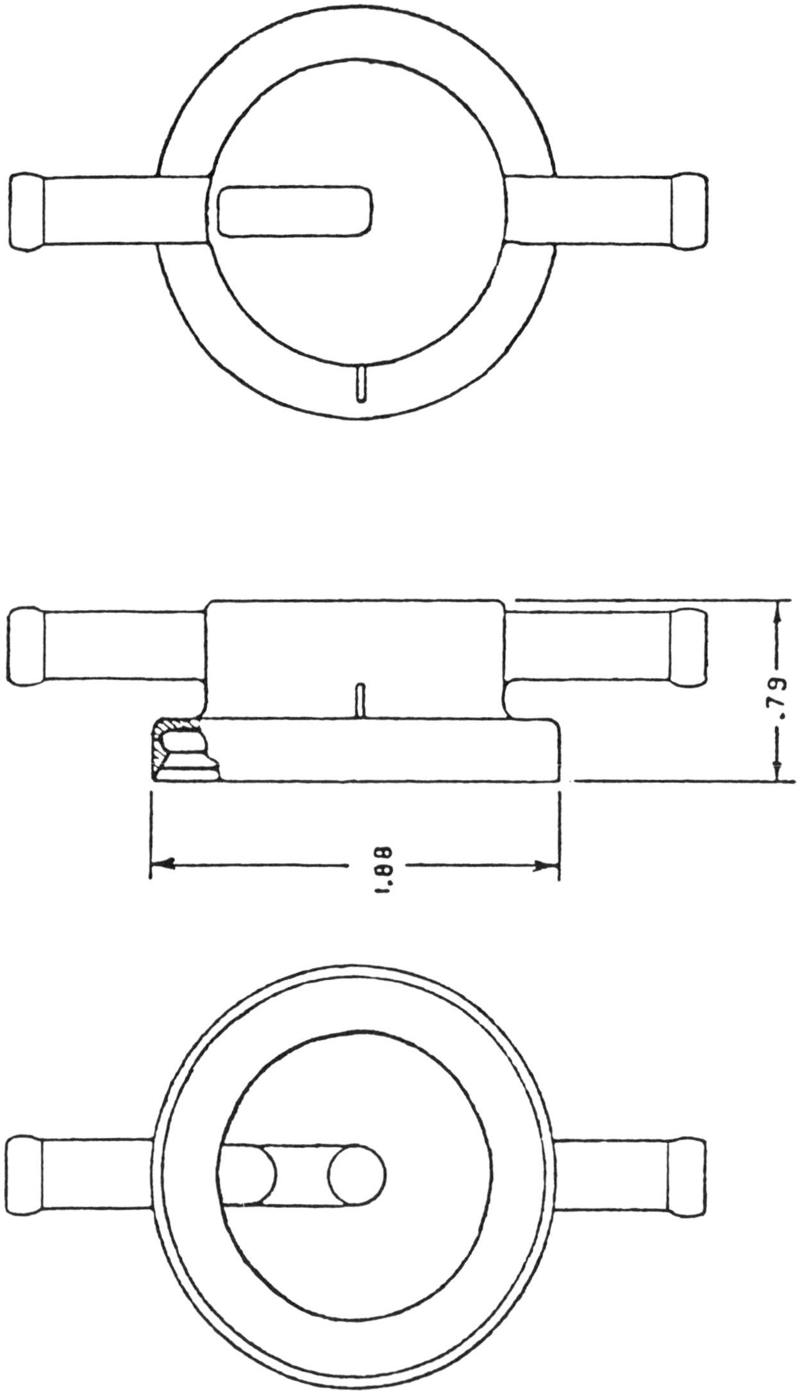

Figure 12. Plastic cap which receives diaphragm.

These two parts are true production parts in a high volume situation. Even though the facility in which they are made and assembled is highly automated, the assembly of the two and placement in the next are performed manually. Both parts are supplied to the operator in separate containers. They are in random orientation in the containers. The separate containers are located on the right and left of the operator. The operator obtains a cap and a diaphragm by simultaneously reaching into the pans on right and left. As both parts are brought forward, several operations are performed. First, the diaphragm is examined and positioned with the proper side up for assembly. Second, the cap is inspected and positioned with proper right and left directions. The two are brought together and the diaphragm is pressed into the cap. At this point, the diaphragm is not completely seated. The operator tamps the diaphragm further into the cap for proper seating in the reentrant lip. The operator then inspects the assembly to assure satisfactory seating, and the assembly is placed in the nest.

All information pertinent to the operations is obtained by tactile sensing, except for the inspection and placement in the nest. Those are done visually.

It has been demonstrated that a Unimate 560 Robot fitted with a simple end effector and using the VAL II language can effectively perform the assembly and placement operations, when coupled with a sensor such as in Figure 2. The arrangement for performing the operations in a laboratory setup is illustrated in Fig. 13. Note that the tactile sensor is securely mounted on the table surface and is a stationary element in the assembly arrangement.

The LTS-100 sensor provides information which enables the following determinations:

A. Sufficiently precise location of the diaphragm

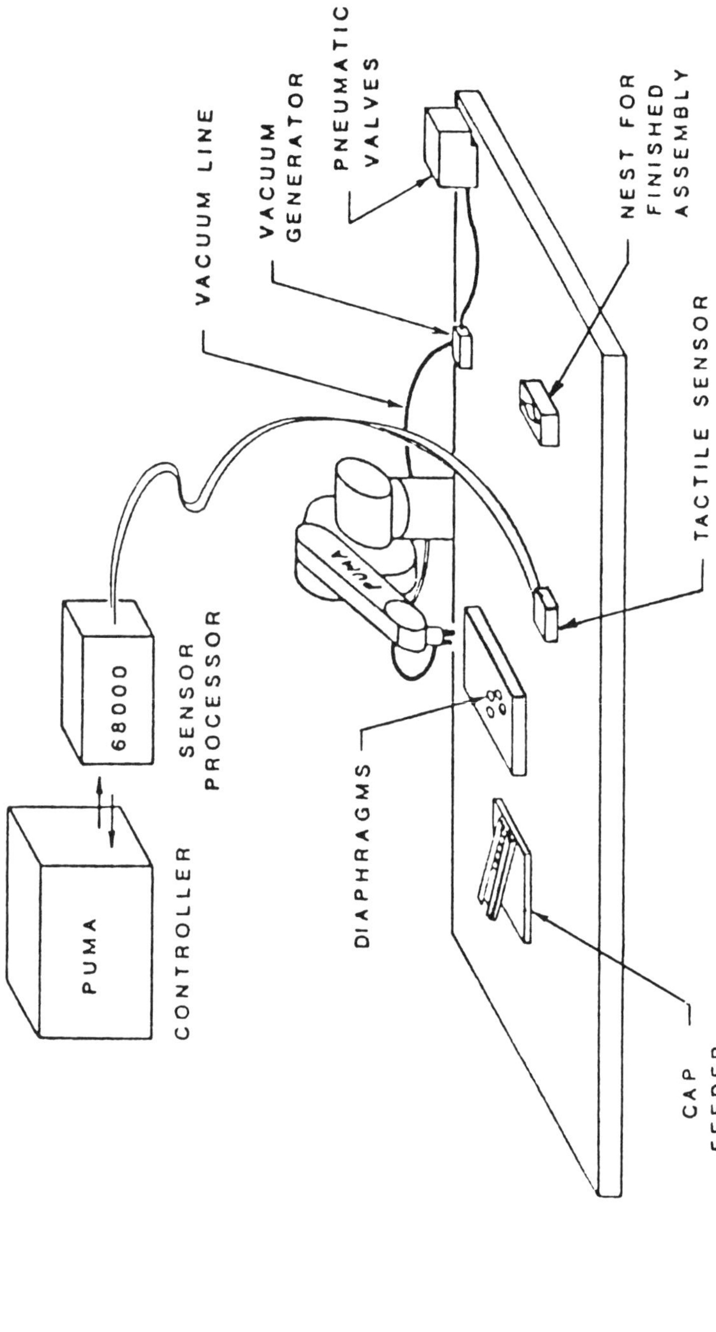

Figure 13. Laboratory arrangement for automatically assembling cap and diaphragm using tactile sensing.

B. Heads or tails orientation of the diaphragm

C. Right or left orientation of the cap

The Unimate 560 controller is configured as the system master with a Motorola 68000-based micro-computer acting as the sensor data processor. The 68000 provides high level results of data processing to the Unimate 560 controller upon request and the controller uses these results in altering the behavior of the robot to perform the task. The assembly process is as follows:

1) A diaphragm is picked from a pile using a vacuum tool and is brought to the touch sensor surface. A centroid reading is taken and the robot moves to this centroid, picks up the diaphragm, and repositions it near the tactile surface center.

2) Four successive sensor readings are taken and assembled into a 16 x 16 matrix. The diaphragm is repositioned between readings in a manner which results in effectively doubling the spatial resolution of the sensor. The centroid of the 16 x 16 matrix is calculated and the robot picks up the diaphragm at its calculated centroid.

3) The diaphragm is positioned such that its central hole is directly over a sensor site and the diaphragm is loaded using the vacuum tool. The magnitude of the reading at this site determines the heads and tails orientation of the diaphragm. The diaphragm is reoriented, if necessary, by dropping it onto the edge of the assembly nest which flips it. The robot recovers it, and proceeds again with step 1 above.

4) The diaphragm is inserted into the plastic cap,

which is presented in one of two possible orienta-ations. A tamping operation is performed to further seat the diaphragm. The assembly is taken to the tactile sensor and the sensor is read, determining the right and left orientation of the assembly.

5) The completed assembly is carried to the nest, being turned appropriately to place it in a final correct orientation.

Textile Issues for Tactile Sensing

Earlier in the chapter, a list was presented which identified characteristics of touching about which taction might provide information. It is useful to repeat them:

1. contact or presence
2. target shape, location, orientation
3. contact area, pressure, pressure distribution
4. force magnitude, location, direction
5. moment magnitude, plane, direction

Although this list was developed originally with workpieces and processes for the metal working industry in mind, the same basic list should be useful in considering the roles of taction in the textile industry. These are generic terms for factors inherent in manipulation actions whether for the metal working or the electronics or the textile industry.

The significant factor which distinguishes the textile industry from the others is the basic nature of the "workpiece". They are made with materials which do not retain their shape, at least in the sense that hardware workpieces do. They are limp. Further, the materials are woven or felted and not homogeneous and anisotropic as metals are. Finally, the component pieces are assembled as plies and not as discrete mating pieces.

These unique aspects of textiles lend interpretations to the items listed initially which are quite different in specifics but generically the same, when compared with other industries. This is a fertile area for study. The introduction of taction based adaptive manipulation is just now being introduced into hardware manufacturing. Much is yet to be learned. Taction in textiles is even further upstream. In fact, we are just beginning to consider its use in textile industry processes.

REFERENCES

1. Harmon, Leon D., "Touch Sensing Technology: A Review," Technical Report No. MSR 80-03, pp. 57, Society of Manufacturing Engineers, 1980.

2. Hill, J.W. and Sword, A.J., "Manipulation Based on Sensor-Directed Control: An Integrated End Effector and Touch Sensing System," Paper presented at 17th Annual Human Factors Society Convention, Washington, DC, Oct. 16-18, 1973.

3. Rebman, J. and Trull, M.W., "A Robust Tactile Sensor for Robot Applications," Proceedings of 1983 International Computers in Engineering Conference, August 1983, by American Society of Mechanical Engineers.

Potential Uses for Robots in the Textile and Allied Industries

Augusto Carroli
Savio S.P.A.

In this chapter, applications of robotics to actual textile processes are discussed by a representative of a company that manufactures both textile machinery and robots. Here the emphasis is on the integration of hard automation (modern textile machinery) with flexible automation (robots).

Mr. Augusto Carroli is a development engineer with Savio, S.P.A. He and his company have been leaders in the application of robots and advanced automation to textile processes.

Potential Uses for Robots in the Textile and Allied Industries

General Information

Staple yarn production follows a process that is subdivided into various phases. Each phase is self-contained and is well identified with the characteristics of the product at a certain intermediate stage (combing, preparation to spinning, spinning, winding, etc.). The constant development and progress of machinery has allowed higher production rates and, consequently, a reduction of the number of machines used in each phase. Improvement of operating conditions has allowed, furthermore, meeting higher quality standards and also more efficient individual processes.

We cannot say the same of the process as a whole, intended as it is to combine a number of discontinuous phases in which specific machinery with different technological characteristics is used. This discontinuity is a result of the different numbers and types of machines used in each phase of each production unit and of the different operational speeds of each phase; it makes it necessary to accumulate the product of each phase in some way, in order to transfer and prepare it for the next phase without degradation. It is clear that this type of process requires much labor, both to carry out the single operations, and to control the production in all its intermediate phases and, at the end, to transfer the product from one phase to the other. Improving these plants will include a considerable employment of automation, which consequently will result in the reduction of personnel for its control and operation. The staff of this new plant will be employed in the organization and programing of the new operating conditions.

In the textile field, the realization of an automated plant will be possible when all the following phases of automation are put into practice, i.e.:

- Machinery automation
- Process automation
- Control systems automation, with output of quality and productivity data.

Machinery automation has to control the operating conditions of a machine, allowing it to run correctly in the absence of an operator. It is usually the first automation step both for machine manufacturers and for users. It refers to all types of devices that are applied to the machine for the purpose of simplifying the process or reducing the labor content (i.e., automatic can and bobbin changers, automatic doffing, stop motions, sliver stops, broken end signals, etc.). This type of automation has already been introduced on a vast scale in the spinning field; nevertheless, it does not directly resolve the second problem of automation introduction-process automation.

The automation of a discontinuous process as in the textile case, is essentially a problem of connecting the different production phases and automatically running the machinery, in order to allow plant operation without manual operations. Consequently the only task of the operators is to control and to intervene in emergency conditions. In other industrial fields, this type of automation is already a reality; in the textile industry the possibility of a relevant introduction is arriving just now, due to recent developments in certain technologies. This introduction makes use of flexible systems based on anthropomorphous or cartesian robots, able to perform different operations in an "intelligent" way, and through "dedicated" systems, which means specially designed devices for work in particular conditions and in conjunction with special types of machinery. In practice, process automation that involves all of the plant will be considered as a final aim and will necessarily pass through different steps, some of which are now being realized.

Automation of control systems and data transmittal is the final step and will allow the execution of specified operating conditions. To conclude, we can say that machinery automation and process automation, realized through an appropriate introduction of flexible and/or dedicated systems, must be introduced without any delay. The results will depend not only on the quality of automation items and on their proper use, but also will be a consequence of a continuous improvement of the machinery efficiency and of the technological development that will be obtained in the future.

Automation in Textile Plants

In order to better understand the possibility of applying robotics to the textile industry, it is convenient to examine problems concerning the different textile operations, starting from combing and preparation for spinning lines.

Automation of Combing and Preparation for Spinning Lines

Combing and preparation for spinning lines has had, in the last few years, the major development, as far as production, versatility and general efficiency of machinery are concerned. Substantial effort by machinery manufacturers has been made to introduce automation devices on machinery in order to increase efficiency and reduce operator interventions. Nevertheless there are operations for which labor is still used, as the changing of cans and/or bobbins, the feeding and discharging of drawing frames, etc.; the high speed of this same machinery requires research aimed at obtaining still higher efficiencies. The next step was to study a complete automation system that reduces or eliminates the necessity to move the cans by hand from one machine to the other and the need for operator intervention to join the broken ends.

After the research and the development of the two-sliver automatic creel that realizes the already mentioned concept of "machinery" automation, Savio was concerned with research and development in the total automation of the preparation process by joining two or more draw frames together. The automatic creel has one or more reserve positions and it is possible to load it continuously from a unique point. All this is of great importance since, while in present creels the cans must be carried by hand close to the proper feeding stations, in this case the same creel, when the can has been loaded in a well defined position, arranges, on its own, to move it to the correct position and to push it aside when discharging is completed. This feeding from a single point allows the realization of a mechanical device that connects the delivery of the prior machine with the feeding station of the automatic creel and that provides for can transport and correct positioning of the sliver.

In a plant with one or more preparation lines arranged in this manner, it is possible to achieve the following advantages:

- Considerable increase of machinery efficiency, since stops and waits caused by interruption of one or more feeding positions are reduced to a minimum or cancelled;

- Drastic reduction of the direct labor necessary to conduct the process since the intervention of operators is limited to emergency cases and to the lot change. (Figure 1)

However, it is obvious that systems of this type could be better used with the introduction of robots and/or of automatic systems suitable for further transport functions. In fact, we know that the labor used to move the packages, i.e., cotton combing laps and cans or bobbins in wool combing lines, amounts to around 30% in Europe. Furthermore this type of labor is onerous both because it is repetitive and boring and because of the physical effort due to the weight of the packages. This labor can be accomplished in two different phases by using either auxiliary mechanical devices that essentially help the operator during the most heavy and boring functions, or by using real flexible automation systems that utilize cartesian robots running on aerial rails to obtain a wide action area on different machines, maintaining low unit costs at the same time. Plants equipped with these automation systems will bring important advantages; less direct labor and a better general efficiency of the machinery. The systems Savio is studying in this field are indicated, for cotton, in Figures 2, 3 and 4.

Another example of the application of the above-mentioned

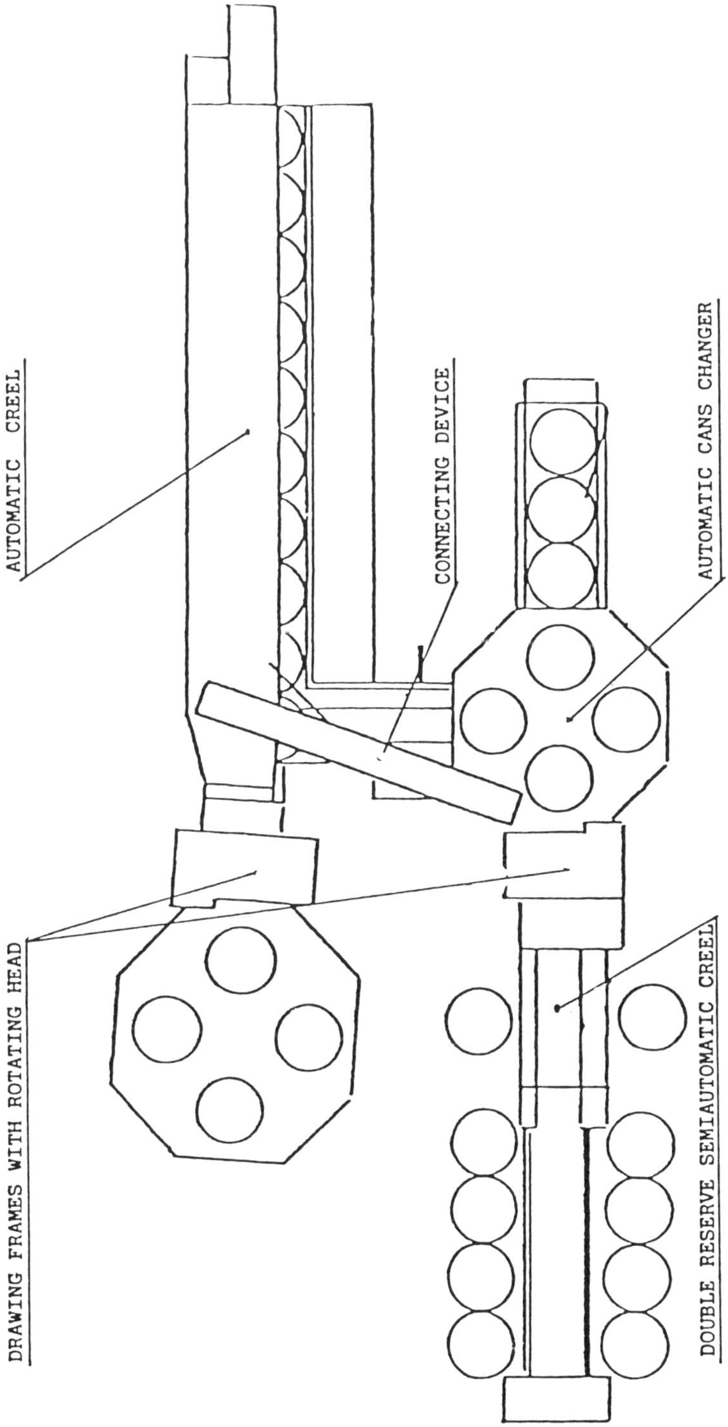

Figure 1. Automatic connection between drawing frame.

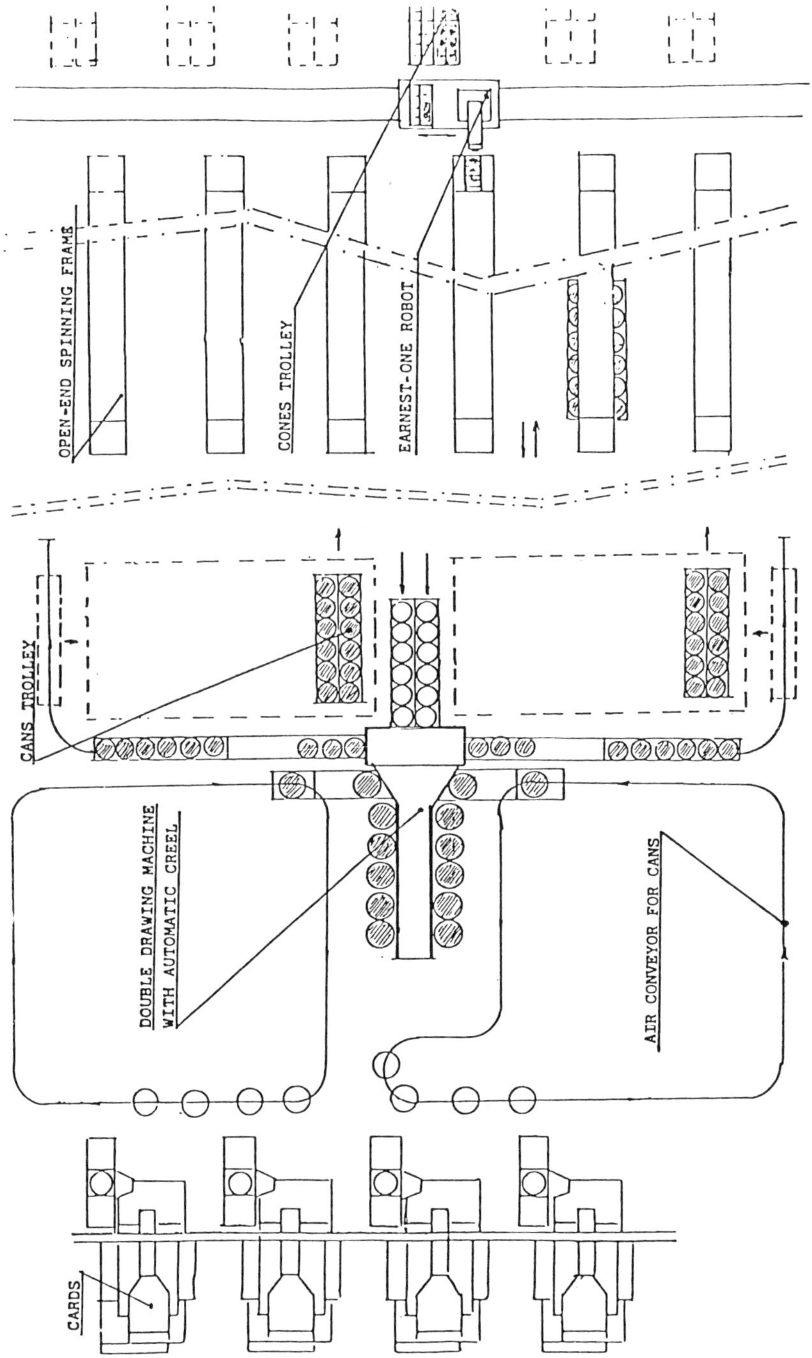

Figure 2. Carded cotton for open-end spinning.

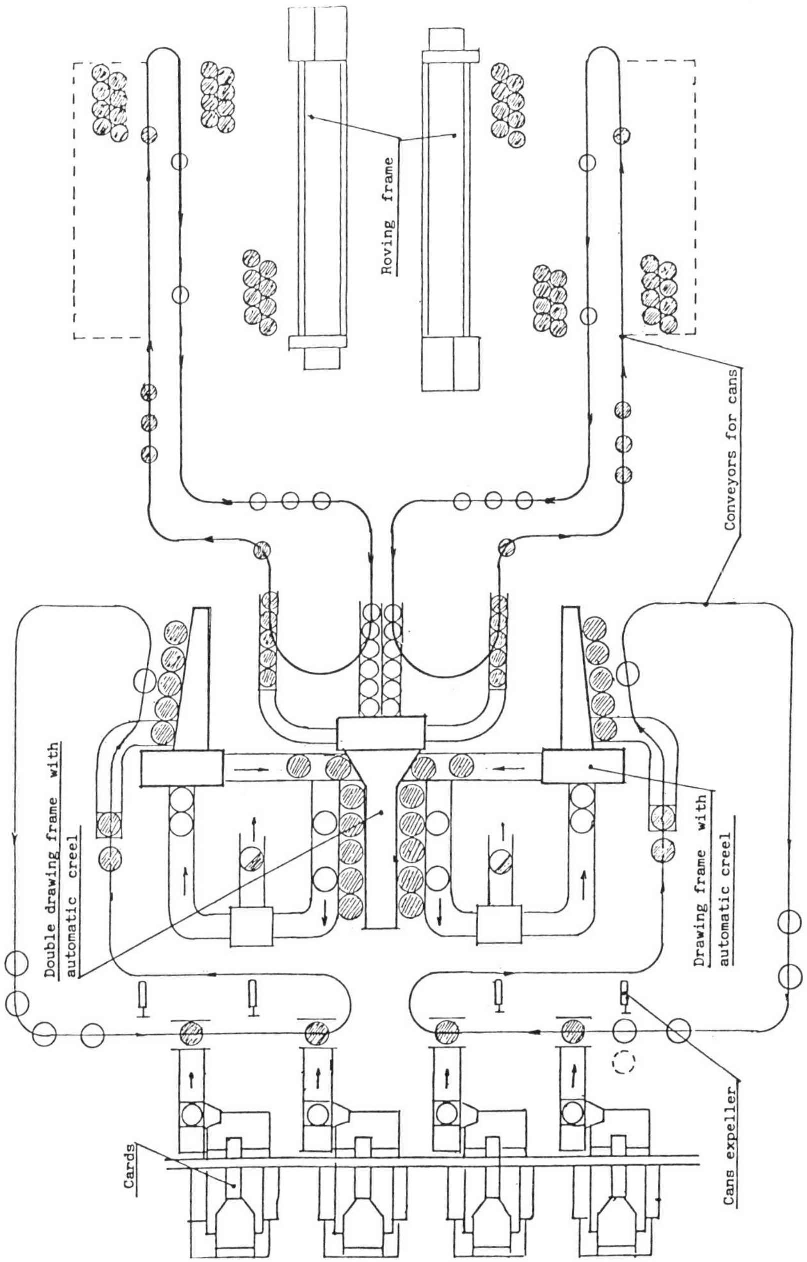

Figure 3. Carded cotton for ring spinning.

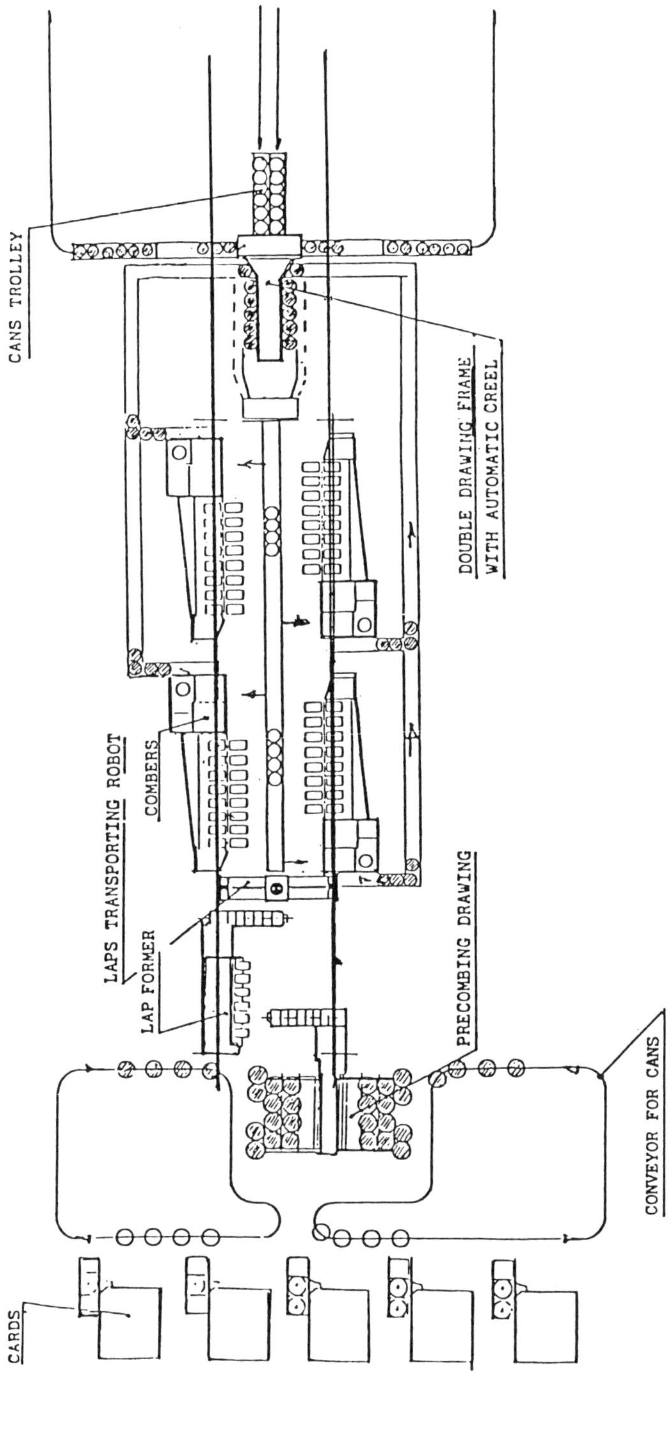

Figure 4. Combed cotton for ring-spinning.

devices can be supplied by a modern wool-combing plant. This type of plant seems to be well adapted for a first automation introduction on an industrial scale. Wool-combing plants are, in fact, generally distributed in big factories; they produce significant quantities of tops with a standardized process and long lots. Labor functions are quite heavy and the machinery efficiency is normally quite low. In Figures 5 and 6 it is possible to see a similar plant equipped with an automation system studied by Savio. Here, particularly, are shown the connection between the before-and-after combing drawing frames, the automatic doffing of the heavy bobbins from the third passage of before-combing and, at last, the automatic palletization and delivery through robots of the bobbins from combing finisher (also equipped with an automatic motion of the trolleys) to the packing and weighing systems. For different wool cycles, in some cases it is possible to form the sliver delivered by the drawing frames, not in bobbins, but inside cans with later ball formation. At the present time this delivery is semi-automatic, which means that before the drawing frame there is an appropriate press with automatic winding which is fed by cans moved by hand. For this specific application, Savio has studied an automatic system for ball formation incorporated in the delivery of machine which provides the pressing, winding and automatic discharging of balls, directly on the draw frame delivery (Figure 7). After this device an automatic palletization and manipulation system is provided.

Automation of Spinning Lines

Among processes for the manufacture of high quality wool and cotton yarns, spinning is probably the process that has had the most recent development and the ring spinning system is the best as far as the finished product is concerned. Without explaining what has been done up to now for the automation of this system, we can say that the automation to be introduced

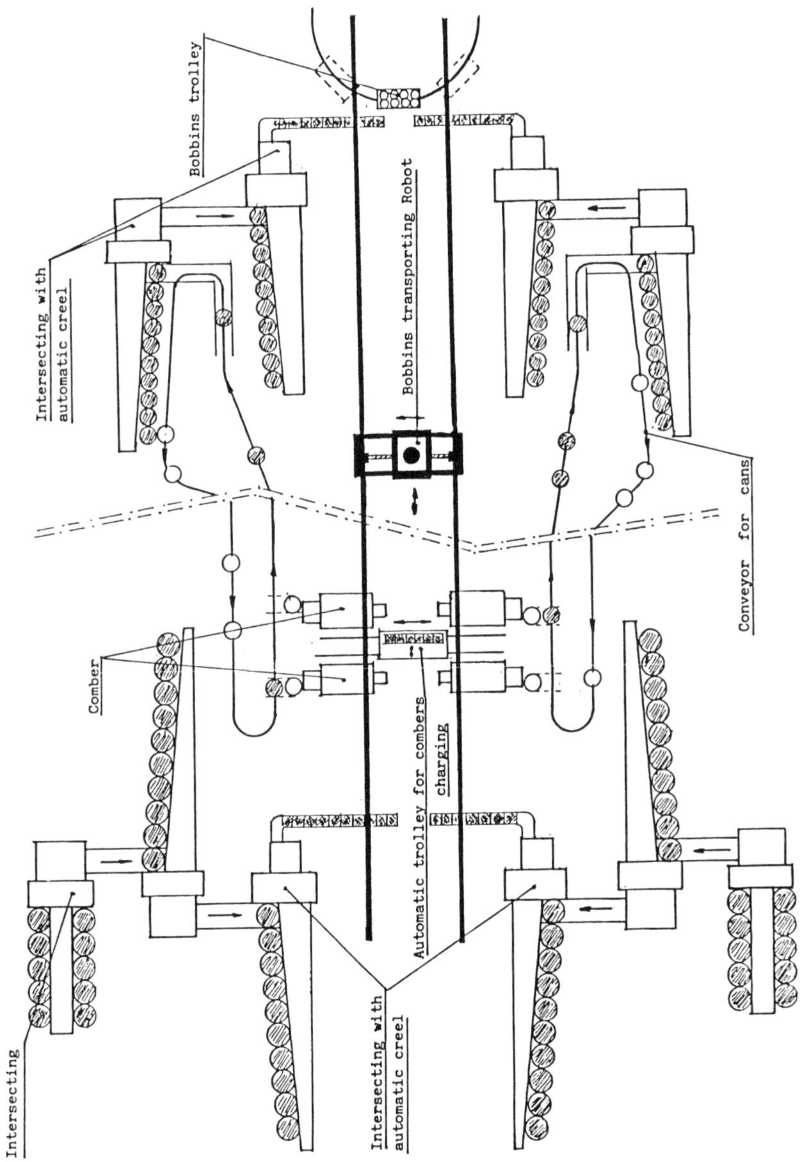

Figure 5. Wool combing process.

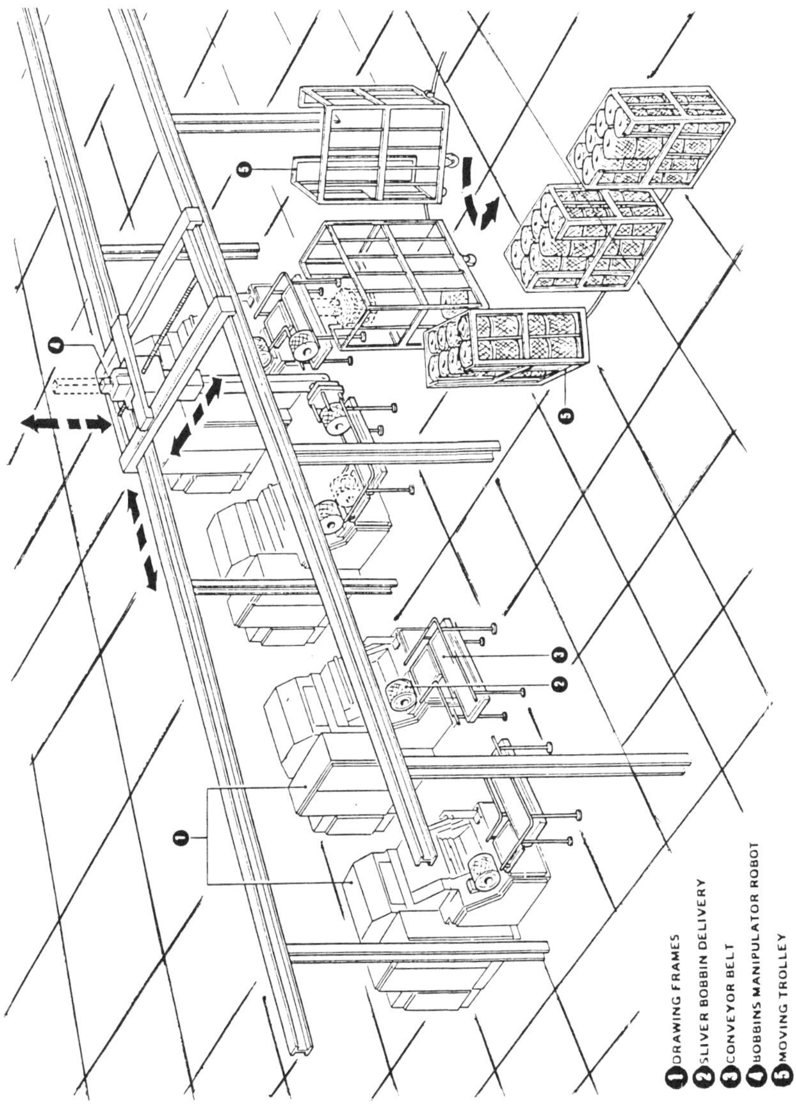

Figure 6. Automation system.

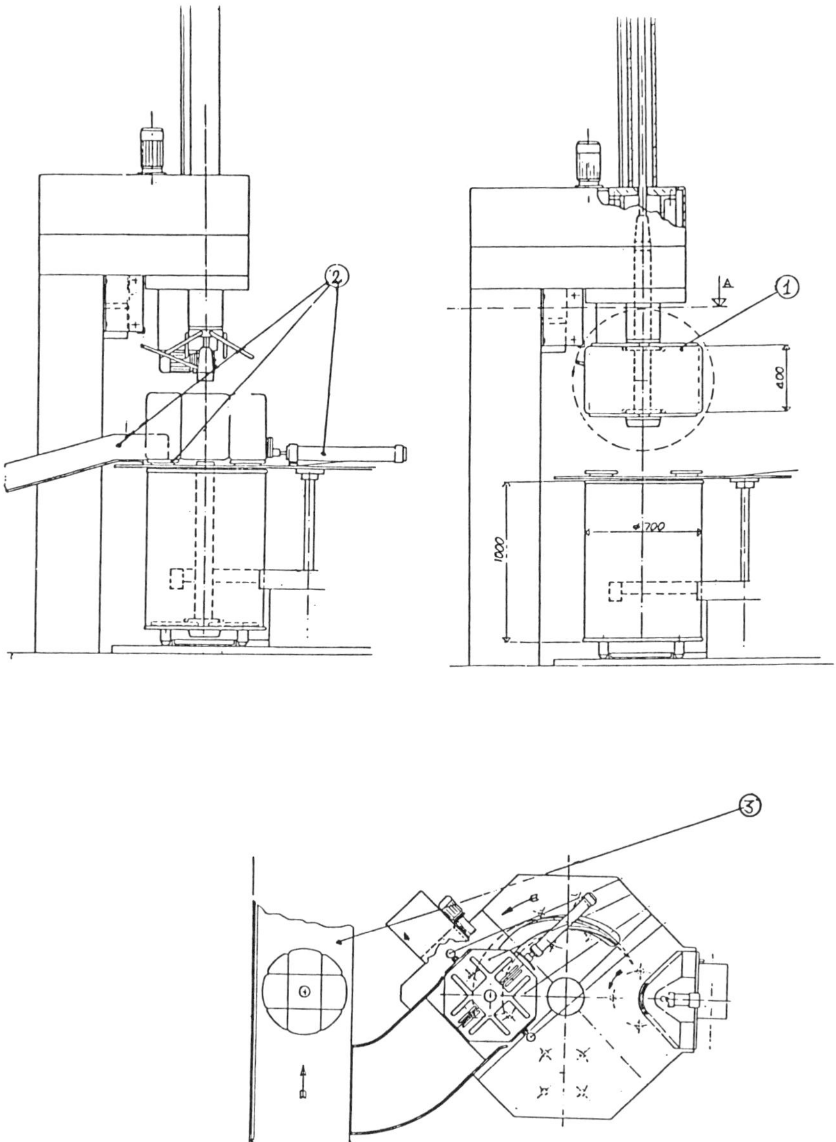

Figure 7. Forming bumps on the drawing frame delivery (1) pressing and fastening bump, (2) expulsion bump, (3) conveyor belt.

into spinning machinery must not only be a "machinery" automation but also a "process" automation. What we have to do to improve spinning automation is, first of all, to apply the concept of the "automatic plant" together with the concept of automatic machinery.

In the area of machinery automation, many developments have been achieved in these last years. Savio and many other manufacturers have studied and developed integrated automatic doffing systems, combined with breakage-control systems, roving feed interruption control systems, and the automatic piecing of broken yarns. Although for particular technical-economic and marketing conditions, the introduction of automatic piecing of broken yarns was not carried out on an industrial scale, automatic doffing and the above mentioned systems had and are still having considerable utilization by yarn manufacturers.

However, in the area of process automation, there still are many things to do, i.e., the connection of the different production phases after spinning preparation: the automatic transfer of bobbins from preparation to spinning frames, the automatic transfer of cones from spinning frames to winders, and the transferring and moving of cones after automatic winding. For these phases, labor in Europe is around 40% of the complete plant labor cost. As far as transferring bobbins to spinning frames is concerned, Savio has studied two alternate systems:

- Robotics system that provides the transfer of bobbins coming from preparation to the spinning frames that afterwards will be loaded on the creel (Figure 8 - 9).

- An integral automatic system that provides, by means of conveyors and bridge crane systems, storage and consequent loading of bobbins in the creel of the spinning frames (Figure 10).

Even if the economic benefits achievable through the first system are quite large, the second one, i.e., the integral one, is certainly more favorable and can also allow the management of small lots.

As far as the introduction of automatic systems is concerned, in order to obtain the connection between the spinning and the winding phases, as well as the winding phase and the following phases, Savio has developed an automatic transfer system between the spinning frame and the automatic winder, according to the development concept of a real integrated machine formed by a spinning and a winding frame where the winding frame is used to absorb the maximum production of the spinning frame. The introduction of the automatic connection between the two machines will allow the achieving of the following advantages:

- Better production and increased quality at winding since the spinning frame cops are not damaged by the automatic transport;

- The high transport capability allows spinning with small cops, which means with a reduced ring diameter. In this way the spindle speed is higher and, at the same time, breakages, wastes and energy consumption are reduced. The splicer solves the knots problem;

- Elimination of transport costs in terms both of direct labor and of means;

- Saving of tubes and their containers and consequently a major economy on investments in plants and accessories;

- Possibility to reduce spinning labor as a result of better productivity.

Figure 11 shows the complete system that includes the spinning frame with automatic doffing, connection device, and automatic winder. Therefore, the system is composed of an integrated machine, in which the bobbins coming from the preparation phase enter the system and the final cones built by the winder come out.

The final automation step in the spinning process involves doffing automation and the successive manipulation of cones coming out of the winder. The system proposed by Savio foresees winder doffing and the palletization of cones on creel trolleys (which can be used either for subsequent processing or sent to packing) by utilizing an anthropomorphous multipurpose robot that has been completely developed by Savio, called "Earnest One" (Figure 12). In this case it is used to serve a number of winders at the same time, since it has the capacity to translate perpendicularly to the axes of the winders.

Further, transport of cones with robots can be studied for the automatic doffing of open-end spinning frames and of double-twisters. In these cases, as for winders, the robot is able to serve several machines on line by translating perpendicularly to the heads where the doffing takes place. The achievable advantages from such a system are of great importance since labor and machinery utilization in the different phases after the winder are quite important, either in the case of additional operating phases after winding in the same plant, or when packing and shipping operations directly follow the winding.

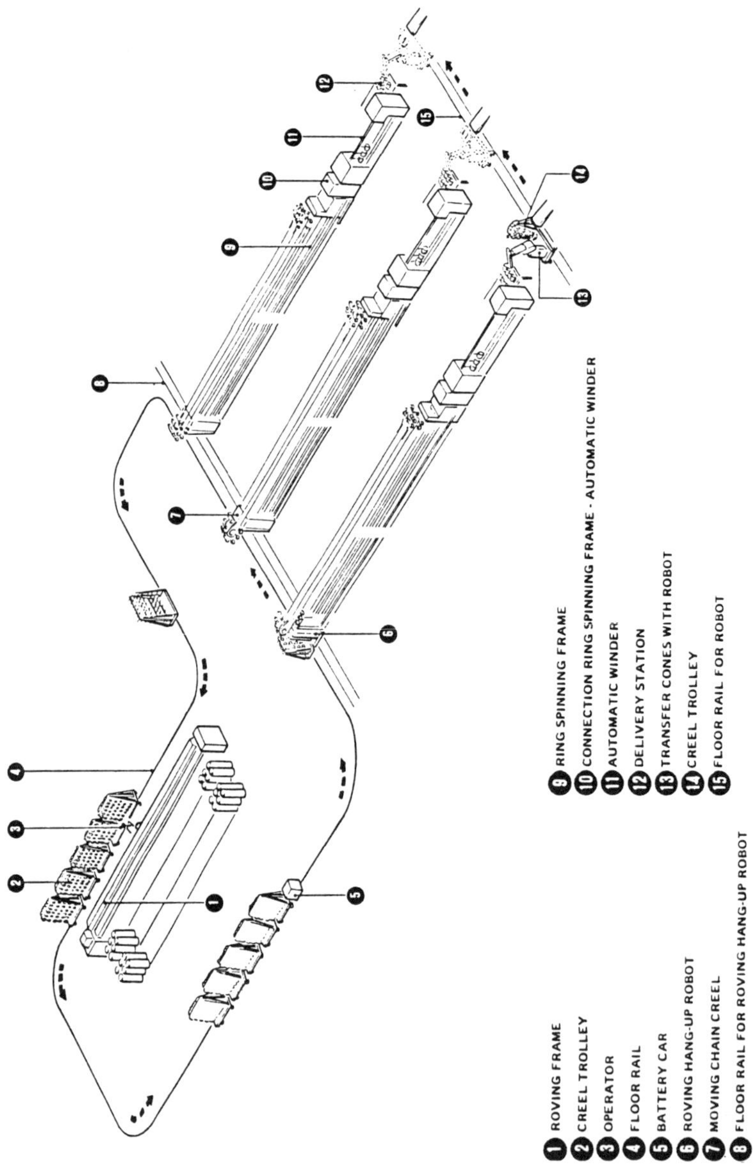

Figure 8. Roving and cones transporting system.

(continued)

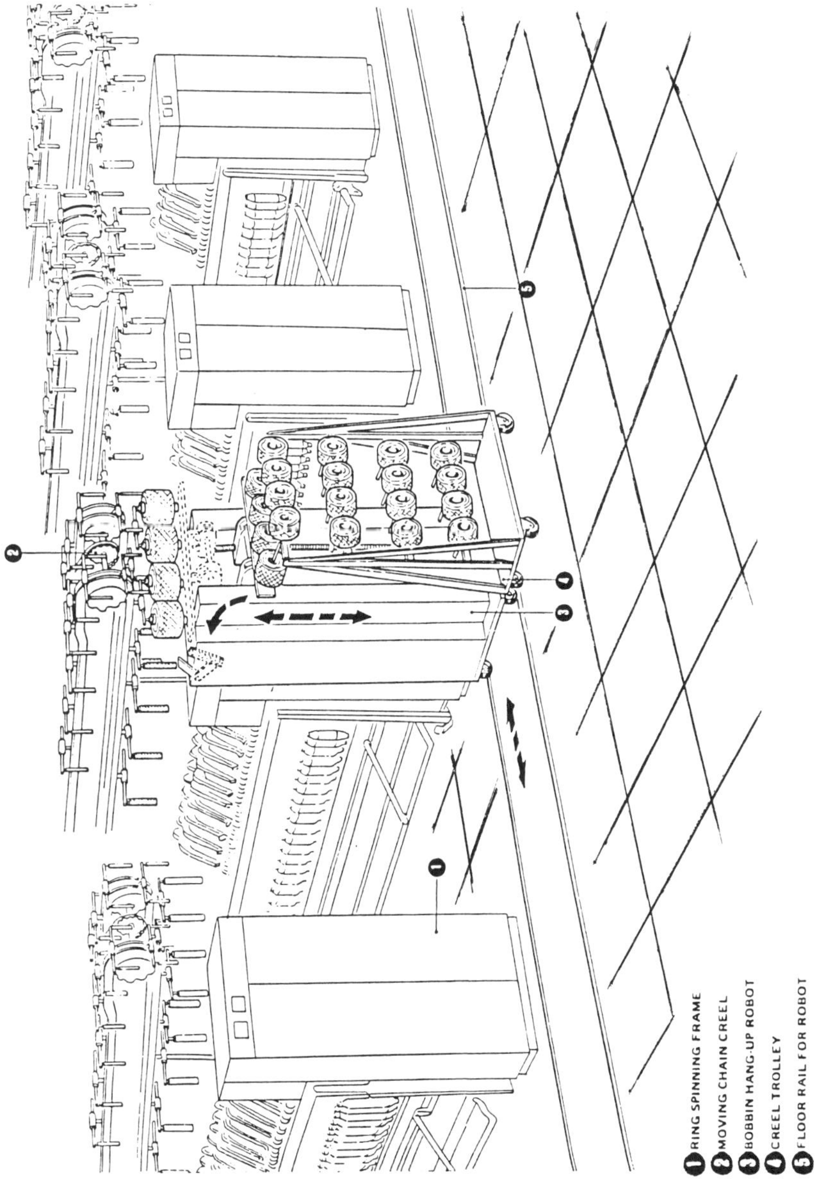

Figure 8. (continued)

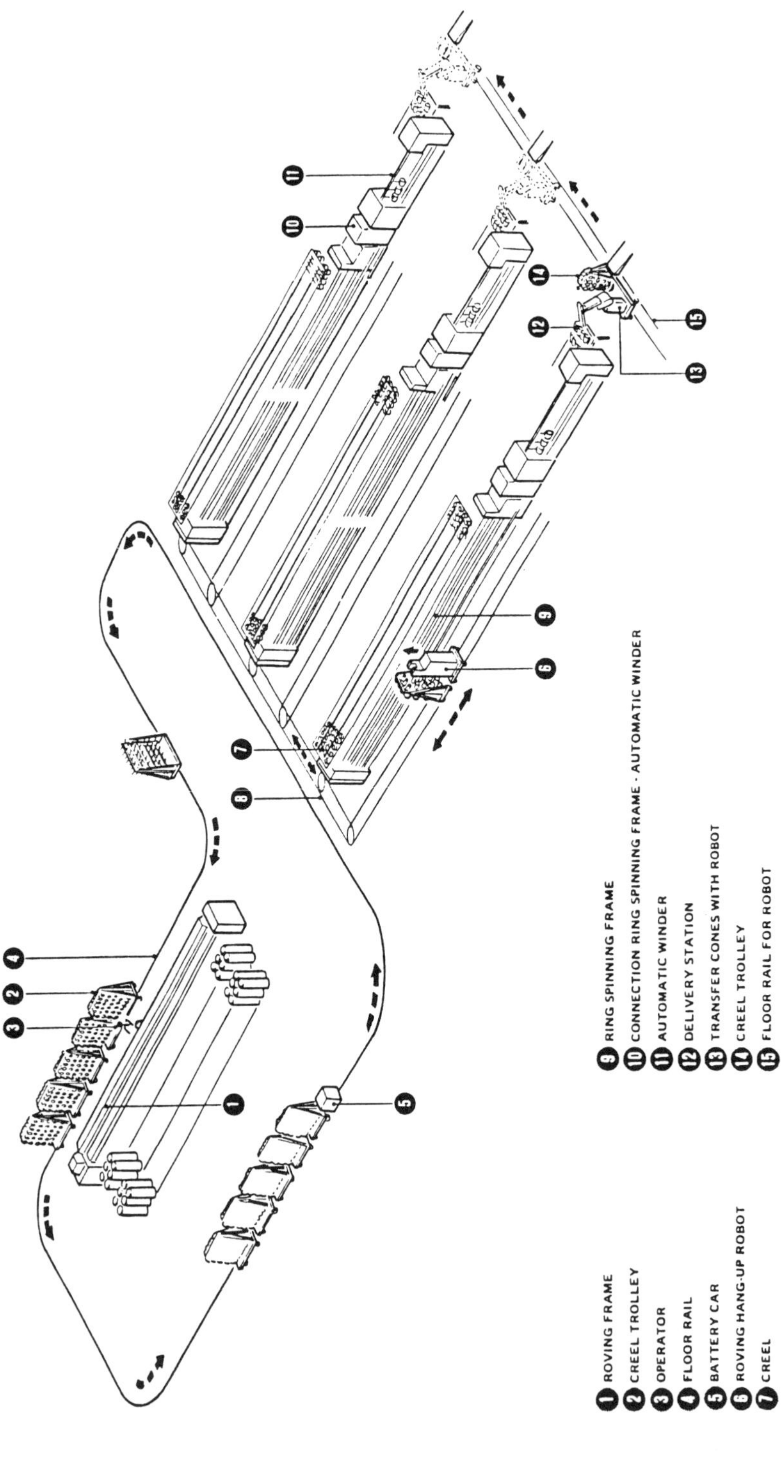

Figure 9. Roving and cones transporting system.

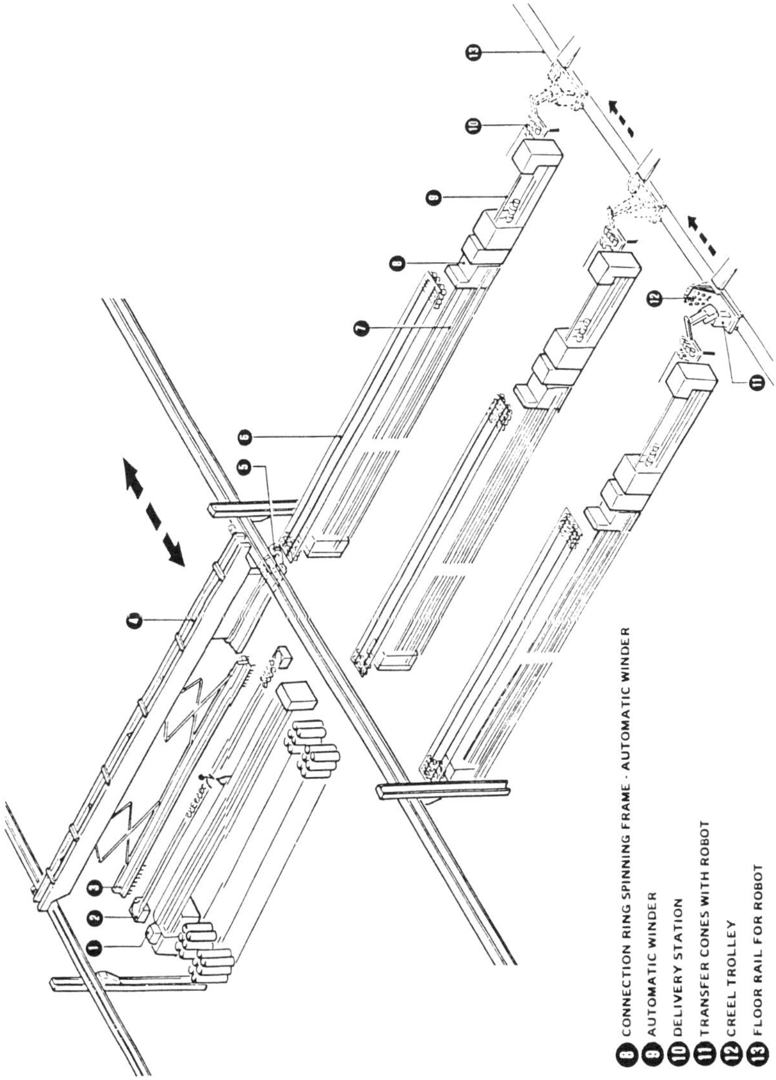

Figure 10. Roving and cones transporting system.

(continued)

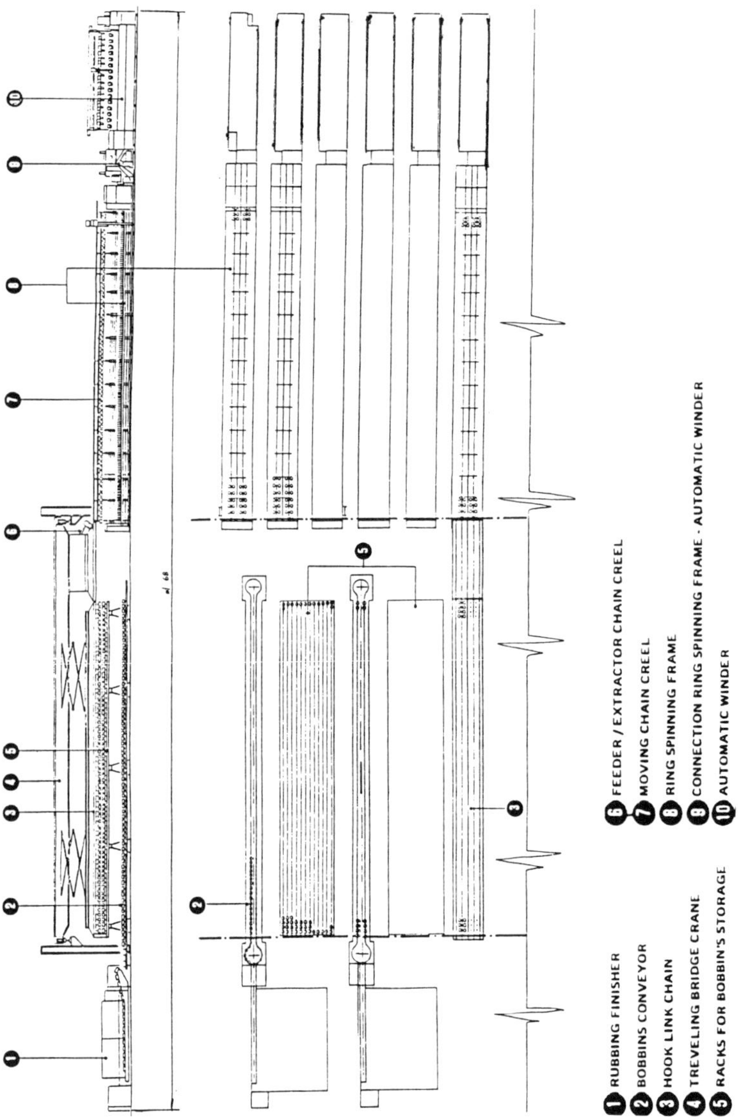

Figure 10. (continued)

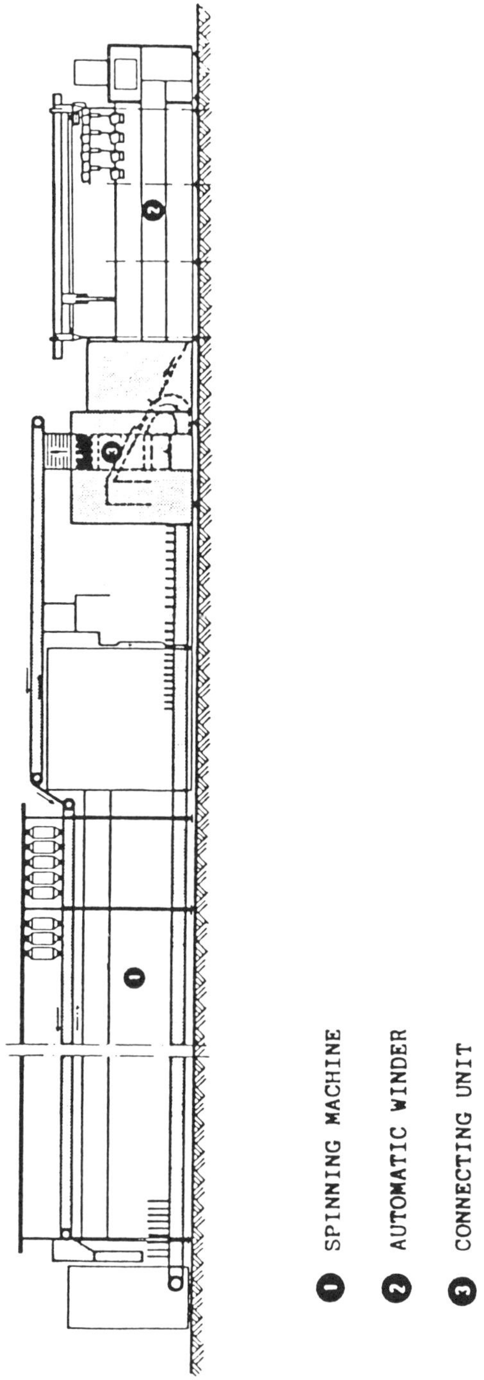

Figure 11. Ring spinning frame with automatic winder.

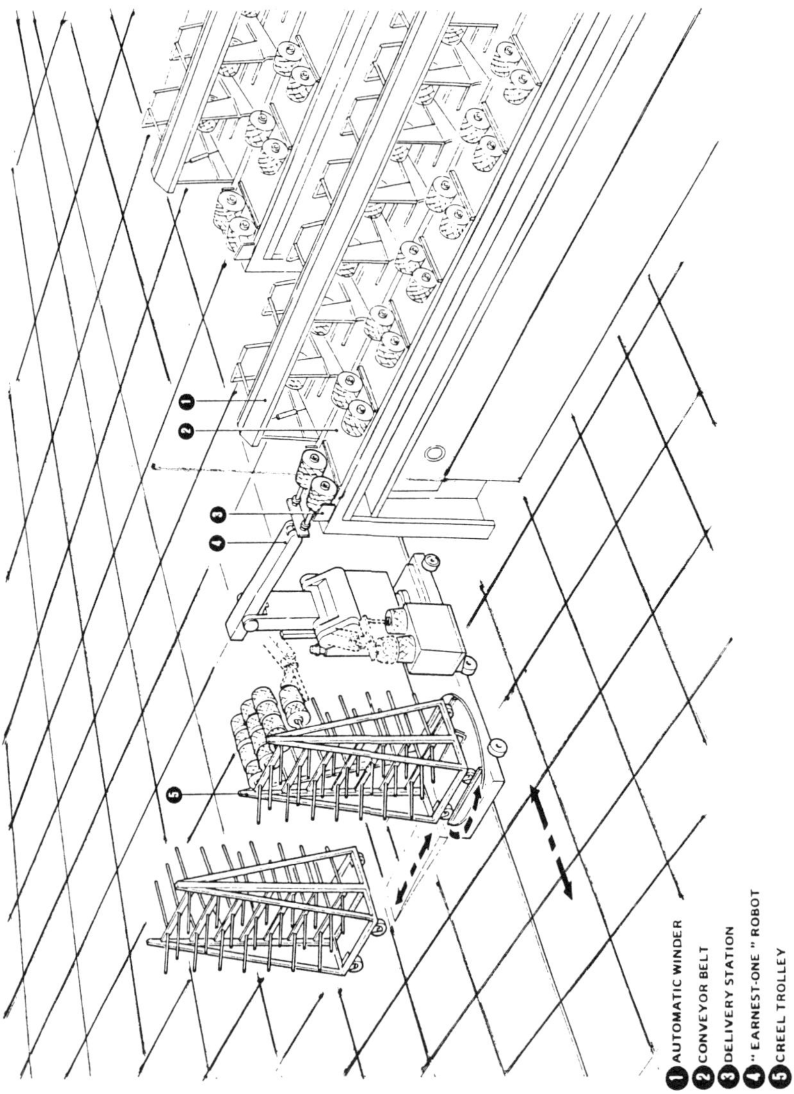

Figure 12. Doffing and manipulation of cones coming out of winder.

Automation of Continuous Yarn Processes

In compliance with the increasing productivity of continuous synthetic yarn processes and in conformity with the leading textile machinery manufacturers, Savio has studied the use of robots for the doffing and manipulation of continuous filament bobbins. The aims of such studies are:

Reduction of operational costs

- Perform in a very short time, different complicated operations during the doffing and string-up of yarns on take-up machines that reach a speed as high as 6000 m/min, with 8 yarns per spindle;
- Reduce the waste production generated during manual bobbin changes;
- Reduce labor on night and holiday shifts for continuous processes.

Improvement of product quality

- Control bobbin weight precisely
- Use computerized systems for the doffing time and the product flows in winding, through storage, to the following processes or to final testing and packaging.

Improvement of working environment

- Alleviate the physical effort imposed by the doffing and manipulation of packages that can reach 25-30 kgs.
- Reduce labor presence in an environment that has become steadily noisier (owing to higher speeds and greater take-up concentrations);
- Substitute for people in repeating and boring operations.

The ROBBIN System is performed by two robots (R1 and R2), moving along the front of the take-up frames. They both utilize the same guiderails under the floor, from which they get electric and pneumatic supply; they are also independent of the bobbin changing operations, and are equipped with relative driving devices and depend on an independent stationary unit only for coordination of space and time. All these operations are conventionally called "bobbin doffing" and consist of taking off all the full bobbins from the take-up units and replacing them with empty tubes, according to a rigid sequence and with well determined intervals. When the two robots R2 and R1 are side by side and have reached the "doffing" position, their operations are (Figure 12):

- R1 unit cuts and sucks the yarns, stops the take-up frame, waits until R2 has discharged the full bobbins (up to 8), replaces the empty tubes and allows the start of the take-up frame;
- Simultaneously R2 goes to the bobbin-doffing station on a creel truck at the head of the take-up line and then to the empty tube dispenser station; the same operations will be repeated for all the take-up frames;
- The independent stationary unit not only coordinates the space and time relations of R1 and R2, but also provides the coordination of bobbin doffing from R2 to the creel; in addition it provides information about the transferred product.

The Savio system performs further important functions: the recovery of the empty tubes, the classification of the product assigned to storage, and to further processes.

Savio, in addition to robots for automatic bobbin doffing, has considered solutions for problem areas after spinning:

storage, feeding and doffing in texturizing and warping operations, and, generally, in product selection and packing.

Possibility of Specific Use of Robotics in Other Textile Industry Fields

So far we have examined the possibility of introducing flexible and dedicated automation in those parts of textile processes that can utilize machinery especially developed by the Savio group. But Savio, in a wider development program of robotics application in the textile industry is interested in, and is taking into consideration, subjects connected with the automation of departments that come after yarn production processes. Some processes well adapted to flexible automation are for instance:

- Tops and cones dyeing, ball manipulation, bobbins and cones including feeding and doffing of autoclave pallets;
- Yarn finishing operations, for moving the packages produced by machinery using these processes;
- Preparation for weaving: feeding and doffing warping creels, where programmability plays an essential role.

Robotics in Textiles

George B. Blomquist
Parkdale Mills

This chapter is a look at the possible future of textile manufacturing by a leader in the application of technology to profitable processes. One of the most interesting points that emerges is that much of the technology necessary to accomplish this future is available today, even though its use is slowed by problems of applications development and cost.

Mr. George Blomquist is a graduate of the North Carolina State University School of Textiles, and currently is Vice-President, Lexington Division, of Parkdale Mills, Incorporated.

Robotics in Textiles

The topic of this paper is decision-making when considering the use of robotics in textile manufacturing, particularly in the sales yarn industry. Before a decision is made concerning the purchase of any capital equipment, it is best to determine, as far as possible, the direction in which an industry is headed. So, first, let us try to visualize possible developments that could take place at some time in the future in a most modern yarn manufacturing plant. This plant would be designed and orchestrated to produce high quality sales yarn at a cost which would not only be competitive for American mills, but also for worldwide competition.

In this modern yarn manufacturing plant, cotton will be received in uncut bales with classing samples attached and contained in moisture-proof bags. These samples will have been removed during the ginning process. In addition to the bill-of-lading, each shipment will list not only the bale number, weight, and grade, but also the fiber test results, such as two and one-half percent span length, fibers shorter than one-half inch, fiber strength, fineness and nearly any other characteristic that may be desired by the buyer. The weights and fiber tests will be so accurate and the shipper will be so reputable that it will not be necessary to grade and sample all bales received. Generally, only a small percentage of bales will be tested in order to verify the quality of the shipment.

Each shipment will be unloaded onto a power conveyor that will transport each bale past a receiving station that will automatically register weights, apply the mill identification number, and remove any samples for testing. All samples of the bales and the shipping lists will be sent to a grading laboratory for observation and to verify the quality of the shipment immediately after it has been received.

After passing the receiving station, each bale will be moved by the power conveyor to a loading station where it will be picked up by a robot and taken to storage. All bales will be stored randomly in racks, each bale individually suspended. The position of all bales with regard to bale number, weight and fiber characteristics will be loaded into a data bank for future determination of each bale's use in the blends that will be necessary to produce optimum quality and high manufacturing efficiencies.

When a bale is selected for processing, it will be removed from the warehouse by a robot on a first in, first out basis. It will be transported to an opening area where ties and bagging will be removed automatically. The bale will be placed in a staging area for blooming prior to entering it into its proper position in the opening line for blending.

Blending and cleaning lines will be similar to those being used today; however, they will run close to 100% of the time and stop only for emergencies. The speed of the feeders will be varied by direct current drives. There will be no more stopping and starting of the lines as is generally being done today. This will provide for more uniform feeding of raw material at a slower production rate in order to open and clean more efficiently.

Before leaving the opening and cleaning area, let us consider the modern handling of salable and reworkable waste. All waste will be collected as is done today, with each description of waste being handled by individual vacuum tanks. Below each vacuum tank will be a container of bale-size proportions. This container will be of sufficient strength to withstand considerable internal pressure. As the level of waste in each container reaches the top, the container will be picked up by a robot and transported to a press, into which the container will be positioned. After the container reaches its proper position, the ram of the press will compress the stock and it will be held in a partially compressed state by retainer dogs. When the ram is withdrawn, the container will be removed by the robot and returned to its original location under the vacuum tank to receive more waste. These steps will be repeated until sufficient waste is in the container to complete a normal size bale. Then, it will be strapped, and bagged if necessary, during its last trip to the press. This procedure will permit efficient handling of all waste and highly automated bale press operations, with only one press normally being required.

Chute feeds that run continuously will feed cards operating in excess of 100 lbs./hour, producing high quality webs. Long-term sliver variation will be so well controlled that the overall coefficient of variation of one-yard lengths will be on the order of two percent or less. This will be accomplished by sensing and controlling the weight being fed and delivered continuously.

With drafting elements located in the calender roll section, manufacturers will be able to spin lower quality yarns directly from card sliver. The cards will be doffed automatically and at random, with specific lengths of sliver. Robots will be programmed to pick up cans from each card, place them onto a sliver truck (eight cans from eight different cards on each truck), and transport them to a staging area for drawing for conventional operation. This will provide for excellent cross-blending and it will be a simple matter for an operator to creel the cans from the truck into a drawing frame.

The drawing frames will run continuously and doff automatically. As previously outlined, one can will be picked up from each delivery for drawing to be placed on trucks which will hold the proper number of cans for the next process. Lap winder laps will be transported to comber creels on conveyors in sections, in order that a complete comber can be creeled with minimum loss of running time. Each succeeding process will be doffed automatically and, where needed, robots will be used to transport full creel sections throughout the remaining processes in the card room.

It will be possible to doff roving frames automatically straight up into spinning creel sections. The spinning creel sections will be transported to staging areas, with the roving description loaded into data banks. Spinning creels will be electronically summoned immediately prior to the time they are needed, transported to the spinning frame sections designated for creel piecing. The length of each spinning frame will be determined not only by ring size and gauge, but also by multiples of the number of spindles on roving frames. For example, for 96 spindle roving frames, spinning creel sections will contain roving bobbins for 96 spindles and sides of spinning frames will be in multiples of 96, e.g. 192, 288, 384, etc.

Spinning frames will run continuously, with each spindle being doffed and pieced up automatically and randomly. This will be accomplished by an automatic piecer-doffer which will be mounted on and will patrol each side of each spinning frame. Each bobbin will run until it becomes full, at which time it will be stopped, doffed, and the end pieced up robotically. This will allow for efficiencies approaching very close to 100% (99.93% for a 30-second cycle time when doffing and piecing bobbins that run 12 hours).

With the low cycle times that will be developed by the random doffing and piecing of spindles, spinning frames will be designed with more narrow gauge and ring sizes - considerably less than two inches for medium count yarns; this will permit much higher front roll speeds than those normally being run today. The doffing and piecing robot will deposit bobbins at the end of the frame into doff boxes which, when either full or as needed, will be picked up by transport robots and taken to the winding area where the yarn will be staged. The description of each spinning doff will be loaded into a data bank and, when needed at winding, the spinning doff once again will be picked up by a robot on a first in, first out basis and transported to a winding loading station. It then will be positioned into a hoist, the doff box will be elevated, and then will become a hopper, feeding yarn to each winder spindle automatically.

The yarn will be cleared while it is being wound, similar to today's practice. However, we will receive much more useful information from the clearing equipment, which will be able to designate different classes of defects either per unit length or per unit weight. In effect, the electronic clearing system will become an effective laboratory tool capable of monitoring virtually all yarn defects on a 100% basis. Most yarn will be wound onto parallel tubes rather than tapered cones because of new developments in creels and storage feeds that do not require tapered packages. This will allow greater machine efficiency both in winding and in knitting, due to the larger packages that can be produced with fewer winding defects than by today's standard for tapered packages.

After winding, all cones will be doffed automatically and transported to an area where they will be packaged by robots, with all cases being strapped, weighed and labeled automatically. All cases will be stored randomly in warehouse areas, with the description of yarn in each carton being loaded into a data bank prior to shipping. When shipping instructions are received, computers will select cartons of each yarn description to be shipped on a first in, first out basis and, through the use of robots, they will be transported to a loading area. In the loading area, entire shipments will be made up prior to the truck's arrival. Each shipment will be encased in portable containers mounted on rail tracks which lead directly to the door of the shipping dock. When the carrier arrives, ridged connections will be made with the transport track to a track which will be built into the bed of each truck. After the truck is positioned at the loading dock, the entire shipment will be loaded and ready for departure within five minutes. This will provide for much more efficient use of manpower and equipment, both in the manufacturing plant and for the carrier.

Computers feeding on-line printers will make up bills-of-lading and invoices which, with the use of communication modems, will be loaded in computer files in the accounting areas of the manufacturer, the carrier, and the customer. No invoices or checks will be mailed. Payments will be made electronically by transferring balances from the customer's bank account to the manufacturer's bank account, with updates on all balances, payments, and receipts being made electronically in each office each day.

Now what I have described to you may seem far-fetched; however, in most instances, the technological art at each point of this hypothetical operation is either being practiced today or is at some stage of development. As you have already recognized, much of the operational and most of the service work done at each process may be handled by robots. The complement of people required to run this plant will be highly skilled technicians and observer/operators whose primary responsibilities will be inspection and correction of malfunctions that may occur.

This is where we come to the most vital and important part of this operation--people. Without properly motivated, trained and skilled people, no operation can be a success. Highly trained employees, skilled in electronics, microprocessors, computers, and enhanced electronic systems will be required to keep this plan operating smoothly and efficiently. All levels of management, in addition to being thoroughly skilled in manufacturing technology and modern management procedures, must also be required either to be skilled or at least thoroughly familiar with electronic and computer systems. Indeed this represents a challenge to all who expect to be successful in the future. With a plant so highly automated, we must not overlook the importance of human requirements. It is imperative that we realize these individuals must not only be highly skilled and self-motivated, but they also must be developed through a quality of life that transcends all fragile relations between man and man, and man and machine.

How soon may we expect to be able to build and successfully operate a plant similar to that just described? It may be sooner than we think! Hopefully, we are in a financial position and are operating successfully in an area which allows us to install capital equipment because we know that it will be necessary in the reasonable future, either because of quality requirements, improving the working atmosphere of employees, cost reduction, or for environmental consideration. Of course, we must not allow ourselves to be lured into a position whereby, with the reduction of direct operators and service employees, we end up with a host of engineers to keep our plant running.

Manufacturers of robots and similar equipment must realize that for their product to operate successfully, all parts, which include both the electronic and mechanical actions, must be thoroughly and absolutely reliable enough to operate on a continuing basis--24 hours each day. Excessive down time on equipment of this type cannot be tolerated. Faults which occur, although they may be caused by high tech difficulties, must be readily repairable by employees who are not so technically qualified.

Personally, I would not want to build and operate a plant today as I have just described it, because I do not believe it would be totally feasible at this point in time. It probably will be practical at some time in the future. Operations in some areas of manufacturing as described are feasible today and each of us with open minds certainly should give full consideration to going ahead and "testing the waters", so to speak, not only to improve quality, productivity, and to reduce cost, but also to begin developing our people and our minds for future events in our business which may be very similar to what has been described.

The use of robots in textile manufacturing will probably

develop in two areas. One will be for those jobs which are laborious and repetitive; the other will be in those operations which make significant improvements in quality and very substantial increases in productivity. Laborious and repetitive jobs which can be handled by robots in the future are those such as receiving and storing cotton, opening bales prior to blending, transport of stock throughout the plant, and packaging. Operating robots which improve quality and productivity will be similar to those used today in automatic splicing on winders and automatic piecing on spinning frames.

Robots are being used in textiles today and substantial developments are already being made. Since we operate in what may generally be described as free enterprise economy, the future use of this new technology will develop as the economics of each area of manufacturing dictate.

No doubt about it, all of us can look forward to exciting developments. Challenges, rather than being reduced, are greater and more enhanced; and certainly, those who succeed will be rewarded by the great future which can be expected for our industry.

Aspects of Robotic Materials Handling in Textiles

T.W. George
North Carolina State University

The thesis is developed in this chapter that technological development in the textile industry has been active and thriving since the Industrial Revolution of the mid-1700's. What seems different in the materials handling revolution of today, which utilizes robots and other flexible automation is more a matter of degree than of kind. The most important problems include redefining the relationship between machines for manufacturing and the workers who are responsible for them.

Professor T. Waller George is a member of faculty of the Department of Textile Engineering and Science of North Carolina State University. His education was in physics at the Universities of Missouri and Illinois. His industrial experience includes associations with the Naval Research Laboratory and Celanese Corporation. His teaching and research interests include fiber extrusion, and the materials science of fibrous materials. He also is active in the Division of University Studies at North Carolina State University, where he participates in programs relating science and society.

Aspects of Robotic Materials Handling in Textiles

The art of textile manufacturing has advanced from the beginning of civilization. It is not surprising that this should have been the case, as life itself was linked to the utility of textiles and it behooved man, at whatever time, to do the best that he could to procure textile materials as he provided for himself the essentials of life and welfare. By the time of the Bronze Age, looms had been invented that utilized various rigid materials, such as wood and horn, to position warp yarns in a more or less stationary configuration so that one could more efficiently insert the filling amongst the warps to produce the woven textile. Whorls and spindles had been invented to assist in the forming of yarn and its packaging. The over-all processes of textile manufacturing had become sufficiently institutionalized to where textile manufacture was a major fireside routine engaged in by members of the family, primarily the females. These textile-making processes involved the transfer of the crude fiber into yarns, initially with the aid of the whorl at the spinning wheel. The yarns then were placed into the hand loom as warps and later onto the bobbins that were loaded in the shuttle, which was thrown backwards and forwards through the yarns of the warp shed to insert the filling into the fabric. The primary power connected with this operation involved the human effort of beating up the yarn thus inserted into the fabric which, in turn, was wound up onto the takeoff device.

As the art of weaving progressed, it soon became apparent that more yarns were needed to supply a loom installation than could be made by a single family. A loom at the hearthside required a large number of daughters constantly engaged in the production of yarn to keep the fabric production from the loom at the level required by the family needs.

This state of affairs obtained largely up to the inception of the Industrial Revolution. There was, however, a brief time when a premature motion towards a factory workplace occurred in ancient Rome. It was associated with the production of woolen goods largely for togas, wherein women from various households met collectively in a workplace to produce yarn and woolen fabric for clothing of the citizens of Rome. This movement was interrupted with the fall of the Empire and it required many hundreds of years to elapse before this early movement towards the factory was to become the generally accepted way of life. The reintroduction of the workplace at the end of the Middle Ages sparked the renewal of the movement towards the modern industrial society.

As the introduction of a factory workplace was renewed by the Industrial Revolution, the factory quickly became associated with a region with easy access to power to run axles, which in turn provided some of the energy for yarn twisting and for loom operation - particularly filling insertion and the beating up of the filling as it was inserted into the warp sheet. The factory provided the requisite space allocation for yarn production so that a balanced production could be achieved. (In France and other parts of Western Europe, spinning halls were introduced for the specific making of yarn, in which young girls were used as laborers. They kept production rates uniform among a large number of spinning positions by timing their wheels as they sang multipart rounds, such as "Pop Goes the Weasel.") As this extra space was provided, so too were mechanisms for the handling of the yarn and conveying it from a point of manufacture to a point of utilization, that is, the point of insertion into the fabric.

While much human labor was still utilized in connection with the early factories of the textile industry, the development of factories occurred in concert with the introduction of nonhuman handling devices wherever technological ingenuity was sufficient. It is this feature of the early textile workplace that contained the germ of automation and robotics as we know the processes today.

By the time that Jacquard looms had developed to the point where one could produce highly-patterned woven tapestries using machinery, the textile manufacturing process had been integrated with what were, in essence, early computers; these machines used punched cards to provide the complex motions that were imposed upon the loom heddles. With this programmed control of weaving machinery, figured tapestries and brocades were obtained which, in some cases, closely approximated the finer handwrought works of art that had been produced primarily in the nunneries of Western Europe.

With the expansion of manufacturing processes immediately

preceding World War II and throughout that period, techniques were perfected for utilizing labor together with advanced machines in the creation of rudimentary "self-thinking" processes. It was a fairly simple matter to devise lathes and grinding machines with human operators for the production of cylindrical shaped objects, such as projectiles and machine screw threads. A simple extension of this process quickly evolved in which the product was formed automatically and then transferred to an inspection station that confirmed the precision of the production. On the basis of a "go/no-go" measurement, the product was either accepted within a given channel or, failing the "go" measurement, transferred into a reject channel. It was a simple matter at that point to automatically transfer the product back to the manufacturing site and subject the item to reshaping if that were possible. Those items which were produced outside appropriate tolerance levels were, of course, rejected permanently.

It was not a difficult matter to extend the go/no-go production inspection scheme to the point where the role of the human operator was progressively minimized to that of an overseer, who ensured that the various automatic setting and conveying steps in the processes were operating properly in their appropriate sequence. From this type of process development, it was a relatively short time until there was developed not merely a go/no-go inspection surveillance of the production, but instruments that could be used to measure chemical and/or mechanical disturbances outside of an accepted tolerance. The measurement was used to activate appropriate mechanisms to move the process back within accepted tolerances. The role of the human was thus removed from the production process, and what many have termed the second industrial revolution was clearly recognized. The second industrial revolution produced devices which replaced not only the muscle of man, as occurred at the beginning of the first industrial revolution but also replaced man's mental capacity and ability to observe with appropriate nonhuman mechanisms, so that now the production essentially was completely automatic. A "generalized robot" had been built to replace man's immediate role in the processes of manufacture.

In the textile industry, one of the first clearcut examples of this total replacement involved the recognition of off-specification product quality. The process was equipped with a means for "feeding-back" that recognition to a station which adjusted the controls so as to obtain the required specification. This early textile process control involving feedback is found in the control of temperature and of size bath concentration, which are used to coat the warp yarns to be utilized in the loom.[1] This type of process control with its inherent feedback and essentially labor-free character was clearly in place by the end of the 1950s. By this time the second industrial revolution was gathering momentum throughout America.

In the textile sector again, semi-automatic production control, developed in small steps in connection with production of synthetic yarns, provides a second example of production with feedback control. This occurred in connection with the control of the viscose dope in the production of viscose filaments.[1]

In the art of filament handling, there are a number of processes that involve the maneuvering of filaments after processing to further enhance the yarn-like character of the filament product. This involves the processes of texturizing which, in general, again involve not only automatic handling, but also a programmable sequence of temperatures that is imposed on the yarn as it is maneuvered through the processing equipment in order to achieve the desired snarled, interlaced, character. Such machines are elaborate robot factories. They not only handle material and call for various thermal treatments of it, but they can be programmed to produce a wide variety of processes, thus giving the textile manufacturer an enormous inventory of yarn-like items made entirely from the synthetic filament technology of the 20th century.

Also in the case of filament yarns, a bundle of filaments appropriately interlaced by means of air jets produces a structure that can be subjected to twisting, plying or air texturizing to produce a looped and whorled structure that can be plied, plaited and/or twisted further. This yields a continuous filament structure that has surface characteristics similar to those of the conventionally spun yarn produced from staple fiber.

Along with these developments of filament technology, newer kinds of spinning processes have evolved, involving conventional staple products. The open-end spinning process involves the introduction of staple fiber by means of an air stream into a yarn forming chamber that imparts intermingling between the filaments and a twisting of the product to produce a yarn which is similar to more conventionally spun yarns. The primary mechanisms in these machines with their robotic characteristics are due to European manufacturers of textile machinery. The beginning of these processes involves suction cups or mechanical fingers that can be moved over the surfaces of the fiber bale to progressively "eat" the material off the bale and convey it through air handling passages to the card. The product of the carding and drawing processes is fed into the open-end spinning machinery.

All of these examples give rise to a certain perspective concerning the current art in the handling of textile materials. It has attempted to provide not only a retrospective review of various aspects of the production of textiles, but also to indicate that, as the history of textile production has unfolded, the introduction of advanced machinery occurred. Ultimately that machinery was such that it performed in many aspects as if it were "thinking", so as to produce a product without the benefit of essential human labor. This is a somewhat more complex view of the robot than that usually discussed. Accordingly, one is forced to consider that a highly articulated mechanical device with appropriate sensing elements to allow it to "see" as well as to "think" perhaps is only a simplification of the more modern production techniques which will evolve in the future.

In the textile industry it is already possible to see places where the robot and the production mechanism may be thoroughly integrated. In the case of certain modern plants the traditional spinning frame is encircled by a traveling device which has articulated motion and is capable of sensing whether or not a given yarn is broken or a bobbin has been filled. At these times the device plucks the filled bobbin and places it on a conveying belt to the production output station or ties up the ends which have broken as a result of a failure in the spinning process.

Alternatively, mechanical devices for conveying yarn packages to looms can be seen in the automatic guided vehicle - essentially small cars moving on a track operated electronically by computers. These are activated when the loom senses that it is about to run out of yarn supply, and an inventory system is tapped automatically. The necessary yarn is placed upon the car, which then is sent in response to the request of the loom station to that loom where the yarn is needed to sustain continued production. A variation of this process, utilizing much of the same equipment, is activated when the loom has produced a full beam of fabric, at which time the loom is doffed and the finished fabric rolled onto the conveying trolley and transported to the production shipping point. This process can be visualized as integrated into a finishing or other subsequent processing plant. All of this more highly integrated production turns the total plant into a complex robot which consumes raw material at its receiving dock in the form of fiber or filament and manipulates it through mechanical conveyance and sensing stations, the results of which are fed into computers which in turn activate additional conveying and moving devices to manipulate the fiber into yarn and ultimately into fabric. The principal human content of all of this is that of the overseer of the process. From the economic point of view, this new labor factor (that of the robot factory) has a fixed cost with a certain maintenance requirement, and the new human labor requirement involved in the production is now that of a more complex reasoner or decision maker who monitors the overall activity of the production unit. The opportunity for this decision maker to engage in entrepreneurial activity is relatively limited, but there still is an essential and major difference in this human labor input, compared with that of the labor that operated factories at the beginning of the Industrial Revolution. The second industrial revolution has, in fact, separated that which is machine-like from that which is intellectually more stimulating from human labor content; it has allowed the human component to perform in a somewhat more elevated environment and transferred the drudgery of the process to that of a nonhuman robot slave.

The techniques for the production of material things underwent a substantial change at the beginning of the first Industrial Revolution. The first revolution can be thought of as the introduction of automatic transfer of power from process unit to process unit conveying work material from machine to machine, or performing automatically particular work tasks utilizing a power source which is available in the workplace.

The second revolution culminates in the recognition of automatic devices which are programmable to stimulate through feedback loops various aspects of this power transmission to accomplish process modifications upstream of the sensing measurement. In this sense the two industrial revolutions can be thought of as end-points of a scale of production which has been continuously varying throughout history from the first toward the second.

The manner in which human labor is utilized has changed substantially. In the beginning it required the collection of workers from their homes into the factory workplace and the organization of these workers within the scheme of production. The second revolution leads to a displacement of many of the worker positions in the production process but introduces the need for more highly skilled service employees to keep the process operating at the desired level.

In the earliest phases of robot introduction into textile processes much of the conventional yarn making and weaving process details are maintained in the newer technological scheme, and the automatic devices with their feedback loops are simply embroidered around the existing processes, thereby reducing the number of human laborers that previously were employed in the chain of production. Devices for the automatic retying of yarn ends on spinning frames where breaks have occurred at particular spindles represents one of the simplest examples of the substitution of automatic equipment for human labor. More recently, robot-like devices with limited articulation have been mounted on carriages which circulate around spinning frames and offer not only automatic retying of broken ends, but also provide doffing and the starting of spinning by the throwing on of yarn onto an empty takeup package. In this way the production of yarn can be made completely automatic from sliver to the wound-up package. It in turn can be conveyed automatically to a next stage in the production chain. Additional robots are under development that will allow the packaged yarn to be picked up from a supply source and conveyed from the point of yarn manufacture to the point of utilization within a weaving or knitting process or the like. The robot transfers the yarn, puts it into the next stage of the textile production chain and starts the production process. These motions require substantial amounts of movement through space and a series of sensing devices that control the subsequent motions.

Another textile-related aspect of the second industrial revolution is found in the elaborate automation involved in the maintenance of material inventories. As an example, a picking device can traverse a predetermined set of fiber bales to produce the necessary tufts and convey them aerodynamically to the card for the formation of sliver which is then fed into spinning devices, for example, open-end spinning processes. The automatic feeding of these devices can be structured in such a way that one can sample statistically the contents of a series of bales so as to achieve the ultimate in blending from a given supply of baled staple fiber. Similarly in the case of continuous filament yarns it is possible to receive at the input of a factory site the large cheeses or straight wound packages, which are transferred to a series of trolleys. When an appropriate encoding is provided at the loading station these convey the yarn to an appropriate spot within the inventory area, to be withdrawn on demand. Such production units are now available wherein hundreds of looms can be found on a given weaving shed floor with relatively small amounts of human labor, on the order of ten workers per 300 looms. The looms are constantly supplied on demand by trolleys working from an inventory containing the necessary yarn components for the fabrics being woven within the factory at a given time. The human labor content is primarily that of correcting various weaving errors which originate from yarn irregularities and providing machine maintenance. In some cases this minimal labor force must sense the approaching need of a given machine for input supplies and signal that need to the inventory delivery system. This type of automatic production of necessity requires either the automatic sensing that a beam has been filled with woven fabric or that the pirn supply for the feeding of the shuttles is running low and requires resupply.

All of the processes briefly alluded to above represent textile processes that have evolved through the first and second industrial revolutions and thereby contain a minimum of human labor. Further, labor which is inherent to the processes tends to be of a rather highly trained and sophisticated variety. Individuals must understand the nature of the automatic sensing devices and the feedback loops, and the controls which are passed through such loops, as the processes for self regulation continue to operate throughout the production cycle. Further the complex mechanical and electrical mechanisms which are involved in this production system involve both elementary mechanisms and highly sophisticated electronic and electromechanical devices. One can quickly infer from this that the processes for programming these processes are indeed complicated and involve a complex set of inputs which can and are changed from time to time as the production is tuned towards higher and higher quality levels or towards greater and greater productivity with some acceptable fault level.

REFERENCE

[1] Zoss and Delahooke, Theory and Applications of Industrial Process Control 1961: Delmar Publishers, Albany, NY.

Sewing Industry Research

Edwin M. McPherson
North Carolina State University

The technological mainstay of the apparel industry has been the formation of sewn seams. For a number of years, research efforts have sought to find a substitute for the sewn seam, but with relatively little success. On the other hand, process automation in the apparel industry depends considerably on either substituting effectively for sewn seams, or on modifying sewing technology significantly. In this chapter, Professor McPherson discusses the status world-wide of research in this area.

Mr. Edwin McPherson has had long and varied experience in the apparel manufacturing industry. After his education at University of the South and Vanderbilt University (MA in Economics and Law), he worked for a number of apparel companies, including Kurt Salmon Associates, H. D. Lee, Kellwood, J. Schoeman, and RCA, and as a consultant to the industry. At this time he is Visiting Associate Professor in the Department of Textile Management and Technology of North Carolina State University, working with the Apparel Management program in the area of modeling studies of current and proposed apparel processes.

Sewing Industry Research*

Introduction

Because of the attention being given to Japanese automation activity, an impression exists that United States sewn products and equipment industries are not active in trying to improve productivity through automation. This is not true. Since the 1960s there have been a variety of attempts to attack both specific and general joining problems.

Joining limp fabric semi-automatically or automatically presents a variety of problems which have been and are a constant challenge to equipment manufacturers and the industry as a whole. It is almost impossible to detail the wide variety

*Reprinted with permission from Bobbin 24 (1) (1), 121-128 (1982)

of attempts which have been made in this area. Basically it appears that the Japanese have proceeded to collect patent information worldwide, organize it and make an orderly assault on creating a sewing system. In making this assault, teams of government technicians, university staffs and industry engineers roamed the world gathering basic sewing system information. A high proportion of what they are studying is of U.S. origin.

Why then haven't the American engineers - mechanics - innovators succeeded in developing a sewing system? What have they been doing all this time? What do they need to achieve success?

Sewing Innovation

Assembling apparel, shoes, tents, curtains, furniture and auto seat covers are but a few of the many activities in which thread or thread substitutes are used to join seams. Joining parachute seams represents a substantial difference in handling techniques than joining a styled outerwear jacket - even though both products may use the same material and type of joining equipment.

As joining activities are investigated, the response to solving problems tends to become product oriented. That is, the problems of assembling a product are fragmented into a series of digestable steps. Ordinarily, production people and industrial engineers solve assembly problems by copying the way they have been previously solved. Equipment firms provide a quick reference point for establishing sewing production lines. When production lines are set up under time restraints, the opportunity for creative innovation is limited.

The sewn products industries have consistently sought to provide better equipment to satisfy both individual and broad industry needs. Modifications of equipment generated by the mechanic/innovator to solve immediate problems are often incorporated into later versions of that equipment by the equipment manufacturer. Handsewing, either actual or with machines, which requires highly skilled operators may be perceived as either a problem or an opportunity to seek solutions. Solutions have tended to create specialized equipment, sometimes of limited general usefulness.

Joining

The Federal Stitch Catalog lists a succession of alternative stitch formations in general use. Machine suppliers compound this complexity by adding a variety of feed and material handling mechanisms. Needle suppliers provide a wide range of alternate shapes and sizes for use on different products and materials. Accessory suppliers display a wide range of folders and button handlers to improve material handling and joining quality. Joining is a highly complex activity which deals not only with many equipment alternatives but also with problems of seam strength, decorative stitching, and in some cases, the creation of air tight seams.

When the complexities of sewn joining are considered in a world of changing fibers, fabrics, wovens, non-wovens, knit and molded goods, it is not surprising that fully automated factories have yet to evolve in the sewn products industries. But, this does not mean that some steps toward a higher level of automation have not occurred. The problem is being studied.

Handling

Perhaps the most frustrating facet of joining parts together evolves from the wide variety of materials utilized in creating sewn products. A substantial portion of existing equipment activity is devoted to developing or improving existing material handling devices. Almost every study proves that moving and positioning materials to, under, and away from the joining process is the major labor time-consumer within the sewn industries.

The Japanese Apparel Automation Project selected thirty material pick-up devices for study in their "System J" project.

Twenty-one of these devices were patented in the USA, five in West Germany, two in Great Britain and two in Japan. From this study, a new technique consisting of a modified pick-up device and "stiffening of material" was developed. What is important in this study, insofar as the American apparel industry is concerned, is that better than two-thirds of all the inventions considered in this report were developed in the United States.

The Japanese preliminary study represents a relatively small portion of all the handling devices which might be utilized for bringing together and positioning parts to be sewn, bonded or otherwise joined. Omitted are overhead and belt conveyors, sensors and switches, indexing equipment and many other devices, in full or partial use within the sewn products industry.

Positioning

Matching one part to another in the joining process is being approached by the machine industry in a variety of ways. Most of the methods use an operator to feed a clamp or clamping point which in turn actuates a series of positioning and joining actions. These semi-automated activities are cam controlled, cam and electronic controlled, or electronic controlled.

Because the methods use equipment to actuate the joining operations, an operator can usually handle more than one machine. Skill levels are reduced while productivity is increased. In some cases a series of activities are combined in one sewing center. Pieces of equipment are combined to do more than one sewing operation or to accomplish intricate jobs semi-automatically.

Joining Displacement

As properties of materials change, the methods of accomplishing product design characteristics also change. While there is nothing new in shaping materials through steaming, pressing, stretching or moulding, the characteristics of many materials allow for a broader use of these techniques today.

Moulding of products ranges from bra cups through wet suits to automobile seat covers. Many of the moulding or shaping techniques involve adhesive supporting materials to improve joining bonds and to improve retention of shape. When these techniques are considered as potential displacement of other sewing activities, another range of complexity faces the designer/production/engineering activity.

Shaping or forming products instead of sewing them has seen broader use over the past 15 years, and is achieving increasing acceptance. Items such as slacks, uniforms and dresses have been formed in whole or in part. Patents have been granted in the United States, Great Britain, West Germany, Japan and Czechoslovakia for making semi-complete and complete garments.

Techniques as varied as spot welding by heating a bow on

lingerie through using sound to cut buttonholes are beginning to achieve acceptance. Blends of natural and manmade fibers offer the innovative designer opportunities to create new looks and better products through the use of moulding and bonding techniques.

Research Environment

Most sewing products firms are small and tend to produce limited volumes of goods for splintered markets. Small firms with limited management are not likely to make investments in equipment research and development. The fragmentation of this industry is, in itself, a barrier to joint equipment manufacturer/user efforts.

The sewn products industries represent a low investment sector of the economy. Sometimes no investment is made in equipment; the manufacturer rents it. In this environment of small, diverse users catering to limited markets, high speed and high volume equipment has a doubtful market.

Sewn products research and development activity is largely conducted by the major sewn products firms. This R&D usually takes one or more of four forms: (1) in-house R&D, (2) in cooperation with an equipment firm, (3) employment of a research firm capable of producing a prototype or (4) employing a university. Two other forms of R&D investigative activity have been sponsored by apparel and equipment firms: (1) the Technical Advisory Committee (TAC) which consists of members of American Apparel Manufacturers Association (AAMA) and (2) the now defunct Apparel Research Foundation (ARF) which consisted of 375 dues-paying members and which received some R&D money from the Government.

In-House Research and Development

Apparel R&D tends to be cyclical and specific. Departments are created and funded when business is good. Projects are normally highly product oriented, usually devoted to either improving quality, reducing cost or both. A high proportion of these projects are basically short term in nature. This form of R&D can often lead to a competitive edge in the market place. Usually, the risk of investing in this type of R&D is relatively low since targets tend to be specific, limited in scope and well planned.

In the earlier example of separating, picking up and positioning materials, 21 of the patents reviewed by the Japanese were U.S. patents. Of these, in-house R&D resulted in patents being obtained by five companies: Stone Mfg. Co.; Cluett Peabody & Co.; Blue Bell Inc.; Russell Mfg. Co.; and Kellwood Co. One patent was held by a research firm, one by an individual, and the rest (14) by equipment firms.

Review of patent records in other areas indicates a similar pattern of activity. It should be kept in mind that several large firms, such as Farah and Haggar, do not usually seek patents yet pursue a reasonably active R&D program. In addition, many smaller firms whose mechanic/innovator develops a device pass the item to an equipment firm to obtain a production item. This increases the equipment firm's patent volume and reduces the activity attributable to the sewn products industries.

Cooperative R&D User/Equipment Firm

This form of R&D again is usually specifically directed at an opportunity. The user is a major prospect for the equipment firm, but ordinarily the equipment firm also perceives a market beyond the user. Projects evolving from user/equipment firm cooperation provide for user testing and user first call on the equipment.

Success in these joint ventures has been spotty, with most of it in the area of combining jobs. Perhaps the best known joint venture is that of Union Special and Cluett/Pfaff in developing semi-automatic assembly of shirts. This venture covered a span of many years and resulted in several innovative machines which are now on the market. There were some ideas which did not work out at that time, primarily because the electronics were not available.

In the recent past, Kellwood/Union Special, Lee/Jet Sew, and Haynes/Union Special provide a sample of firms working together on specific problems ranging from automatic bottom hemming to combining three jobs into one. Both progress and failures resulted. However, most of the projects obtain most of the objectives.

Research Firms for R&D

A number of research firms are in the position to investigate an equipment objective, determine its feasibility and produce a prototype. More often than not, this type of activity has, in one way or another, been jointly financed by several user firms. This approach to R&D was popular during the late 1960s and early 1970s.

As an example, Blue Bell, H.D. Lee and Levi Strauss jointly worked with Battelle Institute on equipment to full-fell jean seams semi-automatically. Some positioning and pulling mechanisms were developed which evolved into work aids, but the basic program of producing a smooth full-felled seam semi-automatically was never solved.

The 375 members of ARF together with the Government commissioned Arthur D. Little Co. to investigate limp material handling devices. This project was successful through the prototype level. The patents were part of the public domain. Unfortunately, equipment producers who move forward with this system to a production model do not have any protection from competitors. Therefore, the results hoped for were not obtained.

Automation Equipment Labs (AEL) approached the sewn products industry seeking backing for a series of automation projects. The firm received quite a bit of backing. The expense associated with attacking a broad spectrum of joining problems developed into more than the backers felt they could afford. As a result, they dropped out of the semi-joint venture.

These examples represent just a few of the joint ventures in which users have attempted to move forward. They do have several things in common:

- The sewn products industries recognize a need for better ways of doing things.
- They have been willing to devote time, manpower and substantial resources to projects.
- There is a limit to how much they can afford to spend on R&D and still stay in business.

Universities

Sewn products firms have supported university research on a limited basis. Most of the industry's support has been directed toward providing equipment and funds supporting courses in apparel activities. Occasionally, advice is sought on chemical or mechanical problems, usually in conjunction with a specific in-house project.

Support of university research tends to derive from individuals within the industry rather than as funded corporate projects.

Technical Advisory Committee

Each year the TAC Committee reviews the state of the art in some area of the industry. Reports are prepared and provided to AAMA members. Basically, the reports are a summation of what the committee has found to be in existence within a specified area. It provides the industry an opportunity to bring itself up-to-date on either equipment or operating systems according to the subject selected for the year. The reports also usually indicate trends and provide management with useful background information.

Apparel Research Foundation

The function of ARF was similar to that of TAC in some respects. During its operation, the foundation collected information on equipment and trends. A quarterly report was issued to its members.

ARF also kept track of government activity which might be of interest to the industry. It obtained funds for some research activity from the government. Among the last proposals it made to the government was one to the President's Domestic Council on October 30, 1971.

This proposal reviewed the state of the apparel industry, the activity of the Japanese apparel industry and proposed a series of counter measures. Among the proposals was a request for funds to develop a preliminary pilot plant design. The proposal suggested that the plant be designated to make military clothing using experimental techniques. The proposal was not accepted. However, the Department of Commerce in 1979 did authorize Georgia Institute of Technology and Philadelphia College of Textiles and Science to investigate clothing and apparel automation opportunities. These reports were completed during July 1980.

Conclusion: Research Environment

An active research environment does exist in the U.S. apparel industry. Sewn products R&D lacks the financial backing to make major equipment and assembly system breakthroughs. Innovation in product assembly will continue to be focussed on limited targets largely using existing techniques. The accumulation of small steps will improve overall industry productivity. It is doubtful that these improvements will occur rapidly enough to prevent further loss of the industry overseas.

Enough is known about sewing systems to warrant experimentation in this area. A tentative single ply assembly system for suit jackets and trousers was designed and described to Hart Schaffner & Marx in 1972. The cost was high and many of the techniques were untested in production. The proposal was not accepted.

A series of proposals for advanced assembly techniques was presented to the Department of Commerce in July 1980. These techniques were accompanied by estimated costs and benefits. Several techniques required technical research. Others involved modifications or combinations of existing hardware. These proposals have not been put into effect.

The cost and risk associated with a major assembly breakthrough preclude the sewn products industry from carrying forward this type of research. If these industries are to grow, then some seed research funds should be provided by government and the industry jointly to appropriate research facilities. In the beginning, development should be encouraged in the high volume apparel areas which are most subject to import activity. As new techniques evolve they should be tested on a variety of lesser volume products.

The thrust of research should be in the direction of bringing into the sewn products industry assembly line techniques. The research can be segmented into manageable pieces. There appears to be every reason to believe that a high proportion of the apparel industry can be at least semi-automated within 10 years under a consistent R&D program.

The Need for Automation

N. Alan Hunter
Celanese Corporation

Is automation really needed in the textile and apparel industries? In this chapter, some of the reasons for the perceived non-competitiveness of the U.S. textile and apparel industries are discussed, and the implications for the survival of the industry are stated. One very important point is that the adoption of advanced technology alone is unlikely to provide a complete solution of the problems facing these industries, without consideration of business and managerial issues.

Mr. N. Alan Hunter was, until recently, Director of Market Relations, Celanese Fibers Operations. Currently, he is on loan to the Crafted with Pride program, an industry program to educate consumers about the importance of buying American-made apparel and other textile products, where he coordinates economics and consultant studies. He is a graduate of Cambridge University in Natural Science.

The Need for Automation

Introduction

In 1983 and 1984, United States imports of garments, apparel fabrics, and yarns and fibers increased dramatically, reaching record levels in virtually every import category. From an import penetration rate of 30% at the beginning of the decade, four years later imports held a 40% share of the U.S. apparel market.

This rapid escalation of imports has brought into sharp focus the precarious future of the industry and highlighted a variety of issues facing the entire fiber, textile, and apparel industry. Among them, the brief period of reduced imports and strong exports experienced in 1979 and 1980 was seen to be very much a function of the weakness of the dollar and only temporary in nature. Second, the U.S. would have to bear the brunt of rapidly strengthening competition from the Far East, which now included the Peoples Republic of China (PRC) as a powerful force in world trade. The European Economic Community (EEC) with its slower economic growth had made it clear during the 1981 Multifiber Arrangement renewal negotiations that they would tolerate no further erosion of their seriously weakened textile/apparel industry. A third realization was that the Republican Administration was essentially Free Trade in outlook, and could not be counted on either to enforce existing import regulations or to display significant new initiatives to prevent disruption of a basic, labor intensive industry.

The sharpened awareness of the problems it faces, both short and long term, has resulted in dramatic attitudinal and functional changes in the industry. Realization of its essential unity is exemplified by the development of AFTAC, the leading trade coalition; growing political sophistication has led to the Labelling Act of 1984, the Textile and Apparel Trade Enforcement Act of 1985 and intensified public relations and lobbying activities; willingness to work together is the basis for such long term projects as the Crafted With Pride Council and $(TC)^2$, the apparel manufacturing automation project underway at the Draper Laboratories.

The purposes of this paper are to discuss the factors that have led to the present, very dangerous, situation, to estimate the future impact of the forces now at work, and to develop, in outline, a technological approach to maintaining a viable, healthy industry. This approach is "pipeline" in orientation and requires no new technology, though it points the way for future developments. It does, however, call for a significant reordering of management priorities, and changes in the way business is conducted. The case will be made that the survival of the fiber/textile/apparel complex is not dependent on legislation, though this would help, but on drastic reductions in the industry pipeline. The rapid provision of high quality fibers, fabrics and garments to the retailer and consumer is the ultimate answer to off-shore supply, which is characterized by long lead times and the uncertainty and risk associated with them, as well as by large, and frequently "hidden", carrying and other costs. Achieving this objective will be difficult, expensive and time-consuming.

The Last Ten Years

In many respects the turning point for the U.S. industry occurred in 1974. By 1973, after a decade of steady growth, the U.S. had reached a per capita consumption of all textiles equivalent to 54 lbs. of fiber, over 45% greater than the average of all developed countries and almost 3 1/2 times the world average (Table 1). It is this disparity in consumption which is at the root of the import problem.

Table 1
Per Capita Consumption of All Fibers
(pounds)

	1964-66(Avg.)	1973	1974	1979-80(Avg.)
Developed Countries	25	37	35	38
- U.S.A.	42	54	48	49
- EEC	28	34	32	37
- Japan	24	44	32	39
Developing Countries	6	7	7	8
- Africa	4	4	4	4
- L. America	9	11	11	12
- Near East	9	11	13	15
- Far East	5	5	5	5
- PRC	4	7	7	8
World Total	12	16	15	16

Source: Food and Agriculture Organization

Following the recession of 1974, which was triggered by a sharp increase in oil prices, the patterns of consumer spending changed significantly. Since 1975-76, consumption of apparel, which accounts for 40% of all fiber consumption, has increased at no greater a rate than the rate of population growth. (Figure 1).

New plants had been designed and built in the early to mid-1970s in anticipation of continued growth. However, the arrival of full maturity for the apparel market inevitably led to corrections being made in the scale and nature of the pipeline's productive capacity as competition intensified at the fiber, fabric, and manufacturing levels. The result was a shake-out of smaller, less cost-competitive products and manufacturers. This process has proved to be a continuing one and it has intensified in recent years, first because of the 1982 recession and later because of the sharp increases in imports in 1983-84.

At the fiber level, approximately 600 MM lbs. of man-made fiber capacity was scrapped between 1979 and 1984 and surplus capacity still exists. In the textile sector, 250 textile mills were closed in the period 1979-84; it is estimated that over 60 mills were closed in 1984 in the Carolinas alone. The impact on the manufacturing sector is much more complex, involving acquisitions, consolidations, and off-shore manufacturing.

The surplus capacity of the soft goods pipeline was matched by that of the retailer. Unlike the production sectors, however, retail floor space has continued to grow, as it has formed an integral part of the retailer competitive strategy. This aspect of the business will be considered in more detail below, but estimates of this surplus range from 50% to over 100%.

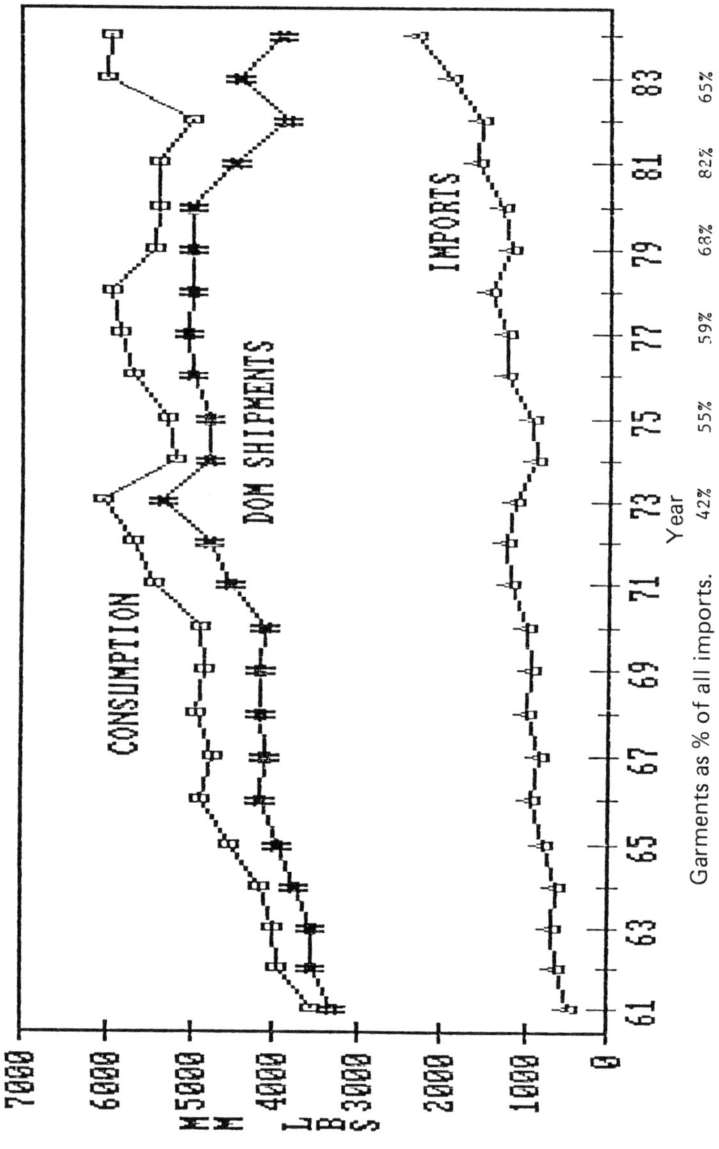

Figure 1. U.S. apparel consumption and imports.

Source: TEB, Dept. of Comm.

Technology, too, played a role in the U.S. textile shake-out. The man-made fiber producers had made major advances in both process and product technology in the late 1960s and early 1970s. The construction of large, low-cost plants led to the dominant position of the U.S. among the world's producers of most man-made fibers. Technological advances hit the textile producers in the 1970s, first in texturing and circular knitting, then in weaving, and most recently in spinning. These changes, which were mainly associated with high speed, low unit-cost processes, called for heavy capital investment that intensified both competition and the shake-out process, as those unwilling or unable to invest either closed or were absorbed by other companies. Though it makes some U.S. mills fully competitive on a world scale, the new technology encouraged the production of long runs of fewer, basic fabrics at a time when diversity is a growing factor of the retail market.

While all of these developments were taking place in the U.S. - and in Western Europe, with different priorities and emphases - a great deal also was happening in the Far East.

Through the 1950s and early 1960s, Japan had dominated international textile and apparel trade, but with increasing competition from Hong Kong. With its ready supply of low-wage labor from Peoples Republic of China immigrants, its highly developed work ethic, its strong customer orientation, its quick response to demand and its textile know-how provided by the ex-Shangai families, Hong Kong supplanted Japan as the leading garment exporter in the 1960s. Hong Kong, although an important fabric producer, was primarily interested in adding the maximum value to fiber and less in developing a fully integrated complex, a policy fully in line with its free port status. Through the 1970s and into the 1980s, Hong Kong has established itself as the quality leader among the low-wage countries.

Taiwan and South Korea had different ambitions. With the strong encouragement and support of government in the 1970s, these two countries rapidly developed fiber, fabric and manufacturing industries on a world scale, dedicated principally to exports. (Table 2). They clearly identified the opportunity to use fabrics and garments as the springboard to broad industrial growth.

Table 2
Exports & Imports of Textile Products
Taiwan & S. Korea

	1973	1983
TAIWAN		
Exports NT $ Billion	47	190
- Apparel	20	118
Imports	18	35
- Apparel	0.5	0.4
S. Korea		
Exports US $ Billion	1.3	6.2
- Apparel	0.7	3.7
Imports	0.6	1.3
- Apparel	+	+

Source: Japan Chemical Fiber Association

With ample low-cost labor available, they started with inexpensive fabrics and garments and gradually moved into broader ranges of higher value products, adding fiber capacity as needed and setting a pattern for other developing countries. It is important to note that the producing capacity introduced in this period by Taiwan and South Korea, particularly for fiber and fabrics was, and continues to be, state of the art, making them cost competitive with anyone. By the end of the decade they had established themselves with Hong Kong as the three largest exporters of textiles and apparel. Currently the Big 3 account for 35% of all United States textile imports and 52% of apparel imports on a square yard equivalent (SYE) basis (Table 3).

Table 3

U.S. Imports of Textiles & Apparel - 1984
(Million SYE)

	All Products	Apparel
World	9785	4722
Taiwan	1335	935
S. Korea	1092	815
Hong Kong	1033	687
PRC	967	445
Japan	737	138
% Big 3	35	52
% Big 5	53	64

Source: Department of Commerce

The increasing dependence of the major Far East exporters on the U.S. is shown in Table 4. For Taiwan and South Korea, 54% of the 10-year increase in man-made apparel exports came to the U.S., compared with 14% to W. Europe.

Table 4

Big 3 Exports of Man-Made Fiber/Apparel

	Taiwan		S.Korea		Hong Kong	
	1973	1983	1973	1983	1977	1983
Total Exports (M Tons)	80	203	78	182	41	47
U.S. (M Tons)	38	102	31	88	10	16
%	48	50	40	48	24	34
W. Europe (M Tons)	16	20	16	44	16	15
%	20	10	21	24	39	32

Source: Japan Chemical Fiber Association

Japan, faced with strong competition, adopted a strategy more suited to its status as a fully developed nation. Taking advantage of the latest technology, concentration on specialty yarns and fabrics allowed it to maintain a strong export position in added value products. As will be discussed below, its objective for the 1990's appears to be to extend technology to garment making, thus circumventing the low wage advantage of its competitors and again competing in world apparel markets.

In 1978, a more liberal PRC government made public an industrial development program that depended heavily on increased exports of textiles and apparel. The targeted production growth was 9% per year on a unit basis. Although reliable data are difficult to obtain, it is believed that the PRC has exceeded this objective. Certainly, the balance of textile trade of the PRC has improved dramatically (Table 5).

Table 5

Textile Trade Balance - PRC

	1981	1984 (9 Mth., Ann.)
Exports	8.4	15.0
- Yarn & Fabrics	4.5	7.6
- Apparel	3.1	5.5
Imports	6.8	3.3
- Yarn & Fabrics	2.3	1.8
- Apparel	+	+

Source: Japan Chemical Fiber Association

The growth of PRC exports to the U.S. has been impressive and is shown in Table 6. To these data should be added an increasing volume of fabric business through Hong Kong and other Far East countries.

Table 6

U.S. Imports of Textile Products From the PRC

	1974	1979	1984
Non-Apparel	77	89	522
Apparel	8	142	445
Total	85	231	967

Source: Department of Commerce

The textile capability of the PRC is clearly enormous and, when coupled with its political clout, makes it a strong contender for the number one exporter position in future years. The PRC's determination to expand its textile operations is exemplified by the number of developments in the last three to four years. Cotton production has been doubled by means of a profit incentive program, making it a net export, and polyester fiber production is following suit. Most recently, a program to decentralize the textile industry will undoubtedly increase productivity, innovation, and competitiveness.

It is appropriate at this point to explore the Far East type of "integrated" industry a little further. Its prime exponents are South Korea and Taiwan but the PRC has similar characteristics, as does Japan.

In the U.S., each element in the pipeline is typically an independent economic unit requiring a return on the assets employed. Thus, as fiber and fabric are sold, the prices include profit margins. In an integrated industry, where exports for hard currency are a prime consideration of the government, there are alternative procedures. The major difference is the transfer of fiber and then fabric at cost, taking the profit needed to give an acceptable return on all of the pipeline assets at the garment level, where the low labor cost provides the greatest advantage. In support of this process, low interest rate loans, low cost raw materials and energy, and advantageous exchange rates are frequently made available. Figure 2 illustrates this process for the U.S. system and for an integrated system. The values shown approximate the costs of producing a man-made fiber slack. The profit margins are sufficient for 20% ROAs on either the individual asset bases or on the pipeline assets. The approach also accounts for the availability of low-cost fiber and fabric in the Far East. As step-wise expansions occur, the surpluses can be moved at cash cost or variable margin plus, as only minimum Sales Administration and Research expenses are incurred.

This example shows clearly that a prime element in competing with Far East producers must be the replacement of labor

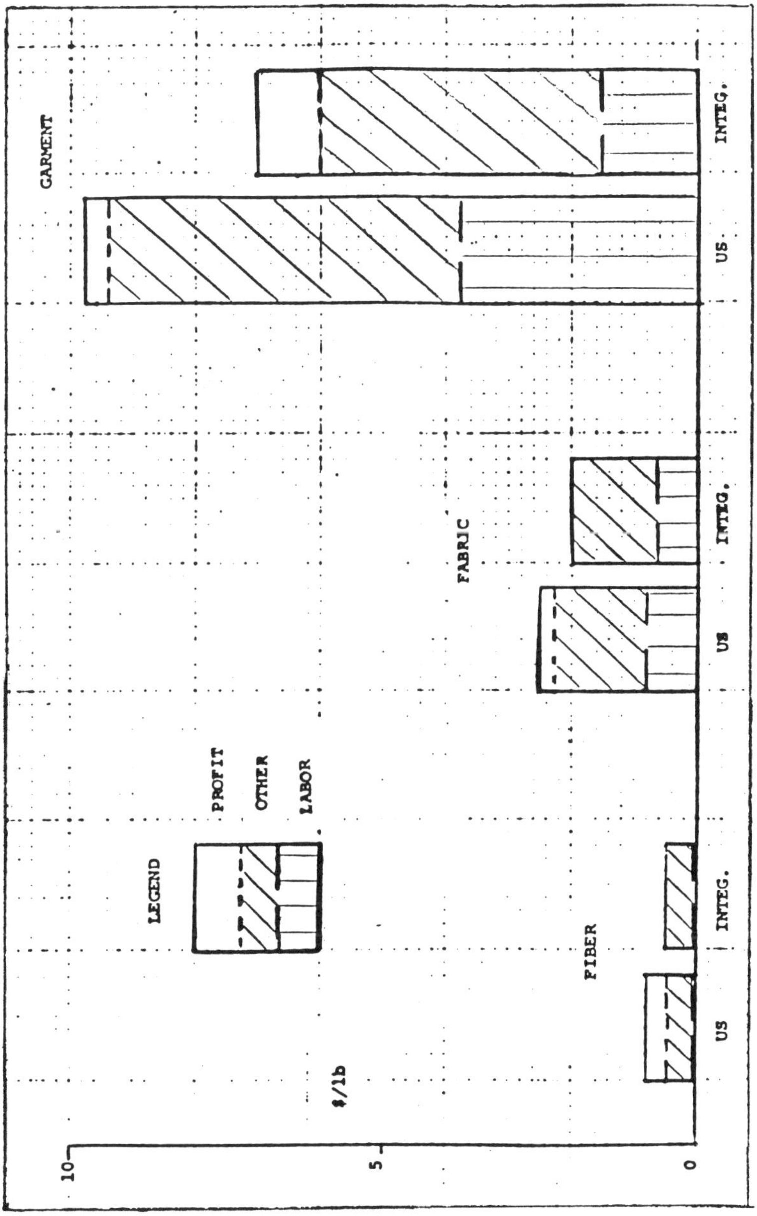

Figure 2. US vs integrated profit patterns.

with capital at the manufacturer level, which traditionally has not been capital intensive.

World Trade Agreements

It is not the purpose of this paper to discuss the general trade policies of the U.S.; however, the surge in the U.S. trade deficit to well in excess of $100 billion in 1984 has focused attention on these policies. There is a growing realization that the trading ground rules used by much of the world are not those originated by GATT or employed by the U.S. This disparity has led to the call for "fair not free" trade.

Many of our trading partners are subsidizing their industries through low cost loans, tax relief, subsidies and other measures to promote export regardless of financial considerations. Other practices proscribed by GATT, such as the use of predatory tariffs, non-tariff barriers, and transshipment of goods to avoid quotas, are also in common use.

It is in vogue among many in government, the universities, and the media to ignore these infractions of trade standards in the name of "free at any price" trading and the need to purchase goods from less-developed countries to prevent international monetary collapse. This argument ignores the fact that the big exporters to the U.S. have little or no bank debt or debt service problems and tend to crowd out the economically weaker nations who do.

The fiber/textile/apparel complex provides an outstanding example of the impact of present policies. Its contribution to the U.S. trade deficit amounted to $16 billion in 1984. In part, this is a function of the strength of the dollar but a second factor is that the other major developed nations are encouraging exports to the U.S. by more tightly controlling their own imports. In 1983, EEC imports increased by only 4-5% versus 25% in the U.S. and Japan decreased its imports by 4-5% through the systematic use of non-tariff barriers. The situation for the period 1979-1983 is shown in Table 7.

Table 7
Growth In Textile & Apparel Imports
EEC vs. USA - 1979-1983

Source	Apparel		Other Textile	
	EEC	USA	EEC	USA
Taiwan	- 6%	+ 59%	-%	+369%
S. Korea	- 13	+ 54	-	+273
Hong Kong	- 11	+ 25	-10	- 4
PRC	+302	+203	+86	+299
EEC		+ 76		+ 71
US	- 60		-62	
World	+ 7%	+ 46%	- 2%	+ 80%

Source: Man-Made Fiber Producers Association

There is another significant difference between the U.S., on the one hand, and the EEC and Japan on the other. It is in the mix of imports. The proportion of finished garments is considerably greater for the U.S.- 50-55% - than for the EEC - approximately 30% - or Japan with 20% or less. This selectivity on the part of the other developed nations has a significant impact on both the trade balance and domestic employment.

The result of these trading patterns was that, in 1983, world trade in textile products increased by only $700 million while U.S. imports grew by $1.3 billion.

The past two to three years have seen the growth of a new import strategem, the use of uncontrolled fibers to circumvent quotas. Silk and ramie are the principal contributors and the sweater market (about 60% imports) is the main target. Table 8 shows this trend.

Table 8

Growth of Uncontrolled Fibers

U.S. Imports By Category:	1982	1983	1984
Uncontrolled	33%	35%	43%
Controlled	67%	65%	57%

Source: Department of Commerce

International trade in textiles is governed by the Multifiber Arrangement (MFA), an instrument of GATT and the successor to the Short Term (STA) and Long Term (LTA) Arrangements Regarding Trade in Cotton Textiles. Under the MFA the U.S. has bilateral agreements with developing nations that spell out allowable import quotas and their growth.

The MFA was drawn up in 1972 with protocols revised in 1977 and 1981. The present term of the Arrangement expires mid-1986. Its primary objectives are:

- to ensure the orderly and equitable development of trade while avoiding disruptive effects in individual markets

- to provide the developing countries with an increasing share of world textile trade

- to achieve the expansion of trade through the reduction of barriers to such trade.

These are worthy objectives and fully supported by the U.S. fiber/textile/apparel complex. However, as discussed above, in 1972 the U.S. textile industry was still in a rapid growth phase, Taiwan and South Korea were still true developing nations, the PRC was largely shut off from world trade and Japan was seen as the principal threat. The apparel import penetration level was 18% and, for all textiles, 9% on a poundage basis. The suggested 6%/year growth opportunity for developing country exports to the developed countries appeared reasonable.

Neither of the Protocols renewing the MFA in 1977 and 1981 fully recognized the major changes that had taken place in the world textile economy. These included not only the growth of the major Far East exporters and their reluctance to open their markets to textile products, except on an as-needed basis, but also their shutting out of the majority of developing countries whose exports to the developed countries have shown little if any growth over the past 10 years. If the PRC and the Big 3 exporters are excluded, U.S. imports from the remaining LDCs have averaged about 25% of all textile imports since 1973. Japan has averaged about the same percentage, while the EEC has averaged 40%.

These factors all have contributed significantly to the inability of the present MFA to achieve its objectives. A second major contributor is the failure of the bilateral agreement mechanism and its administration. Imports from uncontrolled countries and of uncontrolled categories of goods from bilateral countries, transshipment, and fraud all have grown rapidly in the past two years.

The Reagan criteria of December 16, 1983 and the rapid increase in the number of calls have not noticeably affected import activity. The Rules of Origin regulations will probably have a short term impact by closing one type of trade loophole - but others exist or will be developed. A more comprehensive approach to controlling the situation is clearly required if equitable and orderly growth is to be achieved and the viability of a basic, strategic industry is to be preserved through the use of trade agreements.

The Outlook

No change can be expected in the pattern of U.S. demand for apparel in the foreseeable future when compared with the last 10 years. For an average GNP growth, population growth and constant dollar cash income (D.P.I. minus transfer payments minus other labor income, e.g. non-cash benefits) have shown good agreement with actual results over the past 20 years or more.

Other models of imports and exports of garments and apparel-related products project imports - and to a lesser degree exports - to increase over time. The rate of increase in imports is dependent primarily on the value of the U.S. dollar relative to other world currencies but also on U.S. economic growth. Assuming a 3% per year increase in real GNP and a return of the strength of the dollar to the 1972 level, imports are expected to increase from 40% of the U.S. market to 50% in the early 1990's. This projection could well prove conservative; other forecasts made by industry analysts range as high as 70%. This figure is not just speculation; it has already been achieved in the United Kingdom for men's outerwear (Table 9).

Table 9

U. K. Apparel Import Penetration Levels - 1983 - 1984

Men's/Boys' Suits	65%
Men's/Boys' Trousers	83%
Men's/Boys' Shirts	75%
Women's/Girls' Suits	67%
Women's/Girls' Outerwear	42%

Source: Various U.K. Trade Statistics

Analysis of import trends shows that the dollar exchange rate elasticities for both garments and apparel textile products are decreasing. Historically, elasticities have averaged 0.9 for garments and 1.2 for textile products, whereas the current "point" elasticities are 0.8 for both categories. Thus, the return of the dollar to more traditional values will not have as large a beneficial impact on imports as was formerly the case. The decline of both garments' and fabrics' sensitivity to exchange rate fluctuations may be attributed to the importers' learning curve. Once the importer has established and cultivated foreign contacts, both he and the supplier may be less sensitive to exchange rate fluctuations and more interested in maintaining a stable, long-term business relationship.

It should be noted at this point that while models forecast 1983 imports accurately, imports in 1984 exceeded expectations by a significant margin. The over-stocked position of many retailers at the end of 1984 suggests that the level of imports in that year was an aberration and that more restraint will be exercised by importers in 1985 and 1986. Under almost any set of economic assumptions, however, the projected growth rate of imports is positive and this implies a decline in domestic production of apparel.

What, then, of non-economic factors? The forecasts discussed above all rest, implicitly, on the assumption that future political actions or trade agreements will have the same impact as those in the past. However, the overindulgence of importers in 1984 triggered new levels of political and other defensive activities by the industry.

Industry calls for a freeze or rollback of imports culminated in the introduction of the Textile and Apparel Trade Enforcement Act of 1985. The major provision of the Act is that of limiting exports by the major exporters in 1985 to the levels that would have occurred if the provisions of the 1981 MFA had been enforced. In future years, growth of imports from the major exporters would be limited to 1% per year and a control system developed to enforce this rate of increase. Other provisions included more generous access to U.S. markets for the smaller developing nations and extension of controls to include non-MFA fibers such as silk and ramie, which both had contributed significantly to import volumes in 1984.

Passage of the Act into law, even if modified, would have a tremendous impact on the domestic industry, allowing it to plan and finance its future with a confidence not now possible. The greatest, and a very real, danger lies in the enforcement area. Should this be handled without determination, or if exceptions multiply, imports could very quickly become a problem again.

Other political actions affecting imports were the stiffer country of origin regulations announced in August 1984 and the country of origin labelling legislation of December 1984. The former was necessitated by the increase in transshipments of garments and garment parts by off-shore suppliers to avoid import quotas. The latter was an integral part of the Crafted With Pride Council program designed to promote the preferential purchase of U.S. made apparel.

As has already been discussed, the current MFA and associated bilateral agreements have failed to prevent disruption of U.S. industry by imports. The MFA expires mid-1986 and the most probable occurrence is that it will be renewed, even though the GATT and several participating nations, which see it as a barrier to trade liberalization, would prefer it to lapse. It is unlikely that the trading guidelines spelled out in the MFA will be changed significantly - the EEC appears satisfied with the status quo and the emphasis on liberalization by the LDCs will likely off-set any moves by the U.S. and Canada to toughen the restraint mechanisms. The impact of the Trade Enforcement Act on the MFA renewal is difficult to estimate at this time, though its passage would have a pronounced effect on the bilateral process. Without such a modification, the process and its interpretation by Government agencies offers little hope for a meaningful change in the level or rate of increase of imports.

Technological Response To The Import Problem

One aspect of the Crafted With Pride (CWP) campaign has already been mentioned, that of creating a larger demand for U.S.-made garments by the consumer at the point of purchase. This basic component rests on the belief that such demand is essential to preserving the two million jobs provided by the industry. A second aspect is to publicize the fact that, for many garment categories, importing is far costlier than is generally perceived and, when all the less obvious costs are taken into account, there is little to choose between the profitabilities at retail of imports and domestic goods.

The remaining components of the CWP program are more directly the concern of this book and, it is believed, can place the U.S. soft goods industry in the strongest possible competitive position. They center on commercially available technology and are best introduced by outlining the present retail scene.

A major component of apparel retailing strategy has long been to strive for market share by adding floor space. Properly stocked with the right merchandise at the right prices, success should follow. Traditionally, department stores held their dominant position with the help of well promoted branded goods, price points were well maintained and periodic sales events were well controlled and designed to move surplus inventories. The chains controlled the lower price point, volume business but over time gradually moved up-scale, thus increasing market share.

In recent years, the acquisition of national brands and strong promotion of house brands or private label goods by the chains cut further into the department store share and profitability. In the interest of both exclusivity of merchandise and bigger mark-ons department stores relied more heavily on off-shore sourcing and the chains followed.

Unfortunately for this strategy, the discounters followed suit and at the same time intensified competitive pressures by themselves moving up-scale. Figure 3 shows the competitive positions of the major type of retail store. By 1984, 30% of all garments sold were being sourced off-shore by the retailers themselves, by manufacturers, and by importing companies. As exclusivity of design at a realistic price proved harder to find in the Far East, Europe was shopped more heavily and there

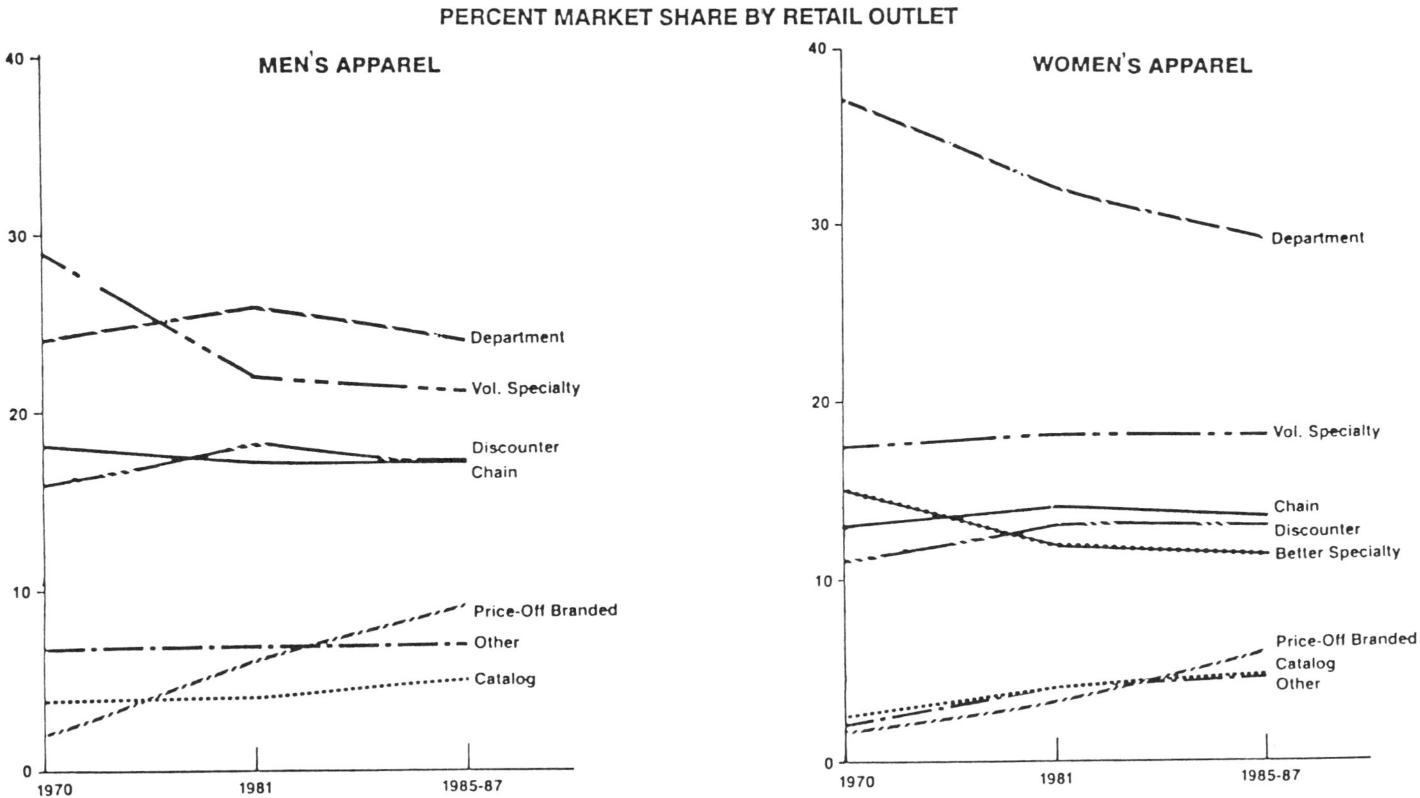

Figure 3. Apparel sales.

Source: Celanese Fibers Operations

the strong dollar was an added inducement.

However, a high level of imports presents its own problems. These include no returns, early payment and long lead times which increase the chance of stock-outs or markdowns. The growing importance of fashion with its attendant unpredictability adds to the problems. The Fall/Holiday 1984 season which ended with an over-inventoried retail trade and heavy markdowns has led many retailers to work on new business strategies, which recognize little or no future growth in apparel consumption and large industry surplus of floor space, but which must cope with an increasingly style - and value - conscious consumer.

The second group of major importers is the manufacturing industry which, without investing more than a token amount of capital, has been able to concentrate on balancing domestic against off-shore manufacture. At home, costs could be competitive for long runs on low labor content garments with good response times, which are attractive to retailers. For short run, labor intensive, or style-rounding lines, imports were the answer. However, here again lead times were a problem, as cutters not only had to carry inventory but also were subject to returns from retailers.

Kurt Salmon Associates has estimated that the cost to the whole soft goods pipeline of long work in-process times and slow supplier response times is of the order of 25 billion dollars per year. This figure includes markdowns and discounts from fiber through retail as well as waste, mistakes, etc. Almost 3/4 of this loss is at retail and the biggest part comes from markdowns. It represents a huge profit opportunity for the industry if the 4-6 months fibers normally take to pass through the pipeline to retail sale can be cut even to 4-6 weeks. It is also a massive deterrent to imports, where the buying process starts 12-15 months ahead of the selling season.

Quick Response (QR) technology is now available to the manufacturer, including CAD-CAM designing, laser cutting, computerized production routing and scheduling, unit production systems, and a large variety of specialized sewing aids. Manufacturers who have installed such QR technology have seen up to ten-fold decreases in elapsed time from fabric received to garments shipped. In addition to productivity gains, reductions in manufacturing cost of up to 20% are possible depending on garment type and complexity. The principal deterrent to installing such systems for the majority of apparel manufacturers is the capital involved. Many believe that significant capital is unobtainable, the pay-back risks are too great, or they do not see themselves as asset managers. Others see their role primarily as marketers rather than as manufacturers.

The driving force for increased use of technology by the manufacturer will probably come from retailers. As yet, they are unaware of the value of QR to the operations in a quantitative sense; yet the profit improvement coming directly from the ability to place orders for goods a week before they are needed, for instance, should not be difficult to estimate. Short lead times not only decrease inventories, but they also reduce stock-outs and markdowns. In addition, they put imports by retailers in a totally new light, drastically reducing or eliminating any cost or service advantage. The retailer who recognizes the opportunities QR offers can be expected to promote the concept forcefully with his suppliers as he, in effect, passes many of his inventory problems to his supplier.

The manufacturer who embarks on this technology path will not be slow to realize at least three things. The first is a reduced incentive to invest off-shore. The second is that importing is less attractive both for cost reasons and because he is now carrying a larger share of the inventory burden. The third and critical realization is that he must pass on his inventory problems to his supplier - the mill or the converter. Such a step will call for QR or Just in Time (JIT) responses by the textile producer and, in turn, by the fiber producer.

As already discussed, the upstream members of the pipeline have invested very large amounts of capital in low cost, high productivity equipment over the past 10-15 years, when flexibility for manufacture of short runs was not a prime consideration. Successful operation in a QR/JIT environment will call for a comprehensive review of alternative ways of doing business and will raise questions, not just of technology but of inventory management and other disciplines. For example, if the pipeline is to be shortened to the minimum, little or no time can be spent checking a supplier's material for quality. Thus, defining specifications and meeting them will become increasingly important functional objectives. So, too, will be an extended flow of information not only about a customer's schedules but also about those of his customers.

Another important consideration is the range and variety of goods to be offered to the manufacturer and retailer. The need for differentiation or exclusivity at the manufacturer or retail level can be expected to increase over time. It would be unfortunate, to say the least, if QR led to increased imports of fibers or fabrics to satisfy private label and similar demands. The provision of wanted fabrics is an important element of the whole QR process.

So far, only currently available technology has been discussed. However, there is a rapidly growing effort to design equipment for the full automation of apparel manufacture. The MITI project in Japan is attempting to adapt and extend robotic techniques used in automobile manufacture to garments and has the end of the decade as the target date for a full scale factory trial. The Draper project of $(TC)^2$ in the U.S. is one of several efforts to automate the majority of operator tasks. In Europe research projects are being written to develop similar technology. The driving forces for this work include not only the replacement of labor by capital and the reduction or elimination of the low wage country advantage, but also the creation of a new industry - high tech apparel manufacturing equipment.

Directions of Technological Change in the Fiber, Textile and Apparel Industries

D.R. Buchanan
North Carolina State University

The perceived directions of future technological change, based on the statements of industry leaders, is the basis of this chapter. Industry leaders, in general, are perceptive about the need for technological change, and generally see it as inevitable, even though they also sense difficulties in its adoption.

Dr. David R. Buchanan received a Ph.D. from The Ohio State University in physical chemistry. After 13 years of industrial experience in the man-made fibers and nonwovens industries, he entered academia and currently is Professor in the Department of Textile Engineering and Science of North Carolina State University. His research interests include the application of advanced automation and image analysis to textile processes.

Directions of Technological Change in the Fiber, Textile and Apparel Industries*

Introduction

Today, every manufacturing industry in the United States is emphasizing the need for technological change in its processes. The fiber, textile, and apparel industries are no exceptions. The reasons why technological change is currently receiving so much attention are the same for nearly all of our manufacturing industries, but they are particularly emphatic for these textile industries. The most important of these, as we all know, is import competition. Only 10 or 15 years ago, imports did not seem to be much of a threat to the United

---

*Based in part on a presention at the 25th Annual Conference, American Association for Textile Technology, Greenville, SC, March 27-29, 1985.

States textile industries, but since that time they have increased at ever greater rates to the point where, in some categories of apparel, some 50-60% of the goods that Americans consume are imported. In past years, an increase in imports did not seem quite so important, since our own industries were managing to keep roughly abreast with exports. However, in the last five years, that situation has changed and today the imbalance between imports and exports is huge. The primary textile industry has been suffering its own import problem and textile manufacturers have, correctly, lamented what seems to be an import situation going out of control. However, the data show that this situation is compounded several-fold for the apparel industry. The problem is made even more severe by the fact that the United States markets for textiles and apparel have reached a plateau. Over the last 15 years or so, the average annual consumption of textiles in all forms has peaked at something in the neighborhood of 60 pounds per person per year. Although there are substantial variations about that figure from year to year the average has remained more or less constant. This means that domestic and imported textile items are competing for the same markets and are not competing in an expanding market. This situation has led to a greater, or perhaps renewed, emphasis on "productivity" in the textile manufacturing industries. It is thought that enhancing productivity will somehow make domestic products more competitive with imported products, which are widely thought to be competitive only because of low labor rates. Sometimes this is true; other times it is not. The import situation also has brought about a renewed discovery that all facets of the textile industries are quite closely interrelated. It is widely understood now that factors that hurt the apparel industry's ability to be competitive also hurt the fiber and primary textile industries as a major supplier. This chain also includes materials, i.e., polymer and fiber producers on the front end, and retail operations on the back end. Fiber producers are also quite aware of the import problems of the apparel industry, but an interesting dichotomy appears in the retail end of the business, in which it sometimes seems that the importation of relatively inexpensive apparel is actively encouraged.

The Nature of Technological Change

What is this technological change on which we collectively pin our hopes for the future? A general term that we hear often is automation. Automation can be classified in two ways: by its technical characteristics and by its effect on the industry.

In general, automation can be characterized according to one of three technological types. The first is mechanization. The term has found uses in most manufacturing industries, but the apparel definition is of particular interest here. Mechanization means, in this context, the building and use of devices that can be described as worker aids, i.e., rather simple mechanical devices usually similar to templates, guides, or other positioning devices that help the operator move material reliably or perform simple operations on that material. Examples in the apparel industry would include a variety of guides to bring fabric to the sewing machine head and, similarly, simple devices to fold fabric reproducibly for certain kinds of seam formation. The second kind of automation with which we should be very familiar in the textile industry is hard automation. Loosely defined, hard automation is identified with machines that are constructed to do one specialist job extremely well. The negative connotation of hard automation is that these machines can do only that one job; to do any other they must be completely torn apart and rebuilt. Examples include many varieties of modern yarn spinning machines, knitting machines, and looms. The third automation classification is intelligent automation; this is the kind that we hear most about today. Its most advanced form probably is the robot, and a robot definition shows exactly what intelligent automation is supposed to be about*. The key concepts in such a definition are flexible, reprogrammable, capable of doing a variety of tasks, and capable of change to do a variety of other tasks.

Automation also can be classified by the effect it has on the industry itself. In this context, we can distinguish between "evolutionary" effects and "revolutionary" effects. Evolutionary effects are those in which a machine effectively replaces a worker on a one to one basis. The machine may perform its job somewhat better than the worker or not, but the key concept is that what is being done is little different than it was before. A nearly perfect example is the technology involved in automatic knot tying in the textile industry. (This example is not quite so good if we consider the emergence of yarn splicing as a substitute for tying; no human could ever splice in the way that modern splicing machines do.) On the other hand, textile technology is full of examples of revolutionary automation--the invention of machines that indeed did revolutionize the way in which textiles were produced. Some examples from yarn technology are particularly appropriate. Using the pre-Industrial Revolution spinning wheel as the starting point, the invention at the Industrial Revolution of a succession of different ways of making staple

---

* The official Robot Institute of America definition of a robot is: "A reprogrammable, multifunctional manipulator designed to move material, parts, tools, or other specialized devices, through variable programmed motions to accomplish a variety of tasks."

yarn--the spinning mule, the spinning jenny, the spindle and flyer mechanism--all of these led to totally new products; therefore, these machines must be considered revolutionary. The modern day descendants of these machines are the open end spinning machine and the various friction twisting machines, which again operate so differently from their predecessors that they must be considered to have revolutionized the yarn-making business.

Historic Examples of Technological Change

To understand what is going on in automation today, it is useful to consider what has happened in the past in the introduction of automation in a variety of industries. As a first example, let us consider hard goods manufacturing, which we can consider to be the making of automobiles, refrigerators, or even electronic components. From the days in which these devices were assembled completely by hand by master craftsmen, to Henry Ford's assembly line concepts, to today's use of the modern assembly cell in which one combines hard automation with robotics, computer control and computer aided design, hard goods manufacturing has progressed steadily through the path of technological change.

We should all be familiar with some of the same progression in the manufacturing of textile products. From the earliest times to the English Industrial Revolution, textiles were produced almost entirely by hand. The only concession to machinery that one might detect was in the area of mechanization or worker aids. Even just pre-Industrial Revolution, for instance as textiles were made in the cottage industry of the United States before the Revolutionary War, whether we are talking about carding, or spinning, or weaving, the equipment used in that era was composed essentially of worker aids, which were used to orient and maintain the position of the parts while a human assembled them into a textile product. However, after the Industrial Revolution and up to the present, substantial and continuous progress in the hard automation era has taken place. The last (1983) ITMA machinery show in Milan showed a number of examples of new hard automation; one of the most striking, perhaps, was a device to automatically and continuously transfer ring-spun packages to a winder, highly automated in the classical hard automation sense. As an example of intelligent automation, we can consider, in addition to robots, automatic guided vehicles which are seen in increasing numbers in textile plants today. These are flexible because they are computer controlled, and they do what they are told when they are told to do it. While the tractor part is completely flexible, the specialized trailer that it tows is an example of hard automation. The result of all of this activity over 200+ years has been that the number of man-hours needed to make a unit amount of a textile product, for instance, staple yarn or woven fabric, has decreased by a factor of 10 every 50 or 60 years since 1760. (Figure 1 [1]). That represents a steady increase in productivity brought about by the fact that new inventions that drastically improve productivity were introduced periodically into the industry. The technologies that seem to hold the most hope for continuing this trend today all are related to computers, micro-processors, robotics, and other forms of flexible automation.

Progress in the apparel manufacturing area seems not to have been nearly so steady. There are few studies in this area and it seems to the casual observer that little progress in manufacturing technology has been made since Elias Howe's invention of the sewing machine in 1846. Many present day apparel plants still rely almost totally on human labor, have confusing paths of material flow, and are operated on a piecework basis. On the other hand, a few plants have a high level of mechanization and hard automation, especially where great

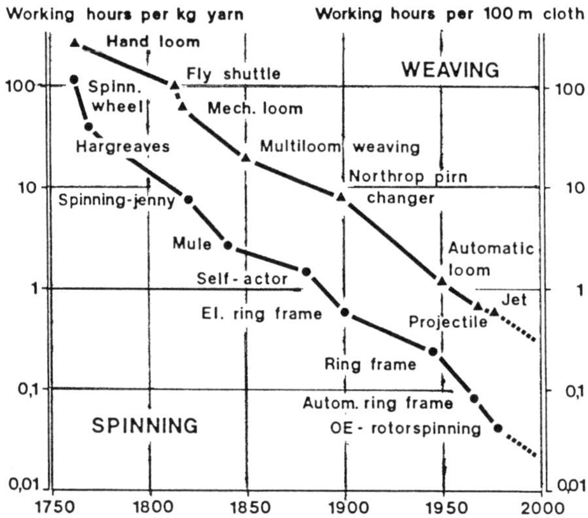

Figure 1. Reduction of human labor in spinning and weaving in the course of time.

Source: A. Heusser, "The Shape of Weaving Machines to Come," *Textile Industries,* 109-112 (1983).

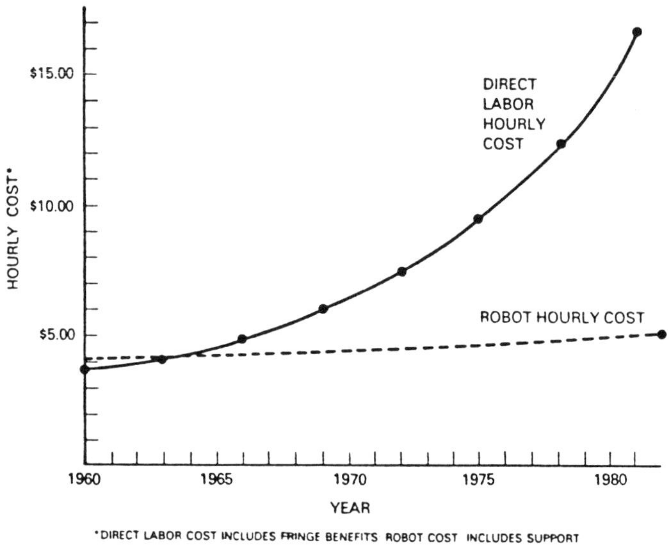

Figure 2. Hourly labor cost of automotive worker versus industrial robot in the United States, 1960-1981.

Source: V.D. Hunt, *Industrial Robotics Handbook,* Industrial Press, New York: 1983.

quantities of one product are manufactured, for instance, denim jeans. Dramatically improving the productivity of the apparel industry may well require utilization of the same concepts that are important for the textile industry--robots, materials handling improvements, vision systems--a whole variety of technologies that are characteristic of those being investigated now in the "Draper" project of advanced automation.

Reasons to Adopt Technological Change

There clearly are reasons to adopt technological change, but they are not always the obvious ones. The reasons usually given fall into three general categories: the reduction of labor costs, the enhancement of quality of product, and the enhancement of flexibility of process. When most of us think about altering labor cost by adopting machinery, we think in terms of replacing workers with machines. And, in fact, this is very often the case. But it also is often the case that this is not the main source of labor savings. The automobile industry, for instance, found out a number of years ago that one of the most compelling reasons for adopting robots for their manufacturing processes and replacing human workers with that technology was not in the current hourly "wage" comparison but rather in the future effects that inflation had on workers salaries versus the future cost of keeping the robot in operation (Figure 2[2]). As another example, the automobile industry has also found that labor costs are effectively reduced simply by virtue of the fact that properly chosen automation can do a job more efficiently. In the automobile industry, a common example is spray painting, in which the main cost savings is the cost of the paint that is wasted [2]. These really are indirect labor costs. A third indirect labor cost is present in the argument that machines can do jobs that people no longer want to do, or that they can do jobs on a continuous basis for which there is a rapid human labor turnover, and therefore very large associated training costs.

Automation can also enhance the quality of a product and there are many examples in the textile industry of hard automation that does exactly that. A good, if arbitrary, example is the card-leveling system, which attempts to ensure a uniformity of sliver that is not possible by any other means. Other examples would include modern weaving and knitting technologies, as well as highly-controlled dyeing and finishing ranges. One of the most important potential benefits of intelligent automation lies in its ability to provide real-time process monitoring, feedback, and control loops, thereby allowing rapid process correction and minimizing quality faults.

Improving the flexibility of a process is a major consideration in deciding whether to automate and, if so, what kind of automation to utilize. IBM has made detailed studies of this issue and has been able to relate, at least qualitatively, the preferred kind of automation to the production volume of one style of item [3]. If production volumes are very low, then human labor is by far the most efficient way to operate a manufacturing process. If production volumes are very high, then hard automation is the most efficient way to operate a process. But for that vast area between very low and very high volumes, intelligent automation or robot is preferred. The practical demonstration of this principle is the huge investment that automobile makers are making in assembly, welding, and painting robots. They must contend with long, but not infinite, runs and there must be model changeovers on a periodic basis. It is said that almost all manufacturing industries are entering an age where they no longer can count on producing the same product day after day, year after year. Flexibility, on a long-term basis at least, is becoming mandatory for any manufacturer of any kind of goods whatsoever.

Driving Technologies for Technological Change

The new technologies that make technological change seem so promising to some and so frightening to others are: computers, robots, and sensing systems. The advent of the inexpensive computer and its internal essential part, the microprocessor, truly has revolutionized the way in which we perform many daily tasks. In manufacturing, there is hardly a plant in the country making anything that does not, at the very least, use computers as monitors of the production process and as a source of management information. Much of our current generation of hard automation incorporates small microprocessors for certain kinds of intelligent operations, and the availability of low cost but very sophisticated microprocessors has made robots a commercial reality.

The idea of the robot is old, but in the past they were too limited in their flexibility to make them very useful. The advent of the low-cost microprocessor allowed robot manufacturers to build increasing amounts of operational flexibility into them to the point where some robots seem "semi-intelligent". Current models of robots are able to do a very large variety of things. These include moving in a number of directions with a number of degrees of freedom, being teachable to do some very complex tasks, being able to react, in a limited way at least, to external events, and having adequate load-carrying capacity and reliability to be suitable for a wide variety of tasks. The technology of robots is developing rapidly and, as it develops, robots become more and more intelligent. Emphasis currently is on: developing a wide variety of sensory systems including vision, tactile sensing, and various kinds of force and torque sensing; giving them much larger memory so that they can remember and do more complex tasks; giving them mobility so that they no longer have to be bolted to the floor in a certain place to do only one job; teaching them to be self-diagnosing of their own ills; and finally, giving them capabilities for inherent safety so that they can operate in environments with humans without danger to either one.

The main effort in sensing systems at this time is in vision. There exist over 100 manufacturers of vision systems [4], many of which are aimed at interfacing with robots to give them the ability to see either two- or three-dimensional objects in black and white or in various shades of gray or, today, even in color. Vision can be monocular or binocular and, in the latter case, gives distance judging capability to the robot. An almost equally rapid area of development is in the tactile sensing area. Devices exist today that allow a robot to pick up and move very fragile objects such as lamp filaments, using just enough force in the gripper to hold the filament without breaking it. Force and torque systems also exist with a common aim of being able to weigh an object as it is being operated on by the robot. This has obvious quality control implications in the textile industry, for instance.

Industry Views of Technological Change

Because the nature of processes in the fiber, textile, and apparel industries is changing so rapidly, and because so many companies with different philosophies are involved, it seemed appropriate to interview some of the people responsible for both the planning and implementation of these changes to determine their views on two fundamental issues.

The first is this: "With respect to current technology, which areas have the most need for improvement and for research?" The second question is equally important: "In approximately ten years, what new technologies will be important and what research should be done now to aid in their development?"

These questions were asked in personal interviews with 27 people in 11 companies, representing the larger and more active companies in the fiber, textile, and apparel industries. The respondents included vice-presidents and directors of manufacturing (7), vice-presidents and directors of research and development (9), vice-presidents and directors of engineering (4), and seven others with diverse titles and job responsibilities, such as purchasing, personnel, product groups, and industrial engineering. These responses indicate the probable directions of technological change in the near future for the fiber, textile, and apparel industries [5, 6].

Current Status: The Fiber Industry

With its technological roots in the chemical process industry, the man-made fiber industry in this country has become quite sophisticated in the areas of monitoring, feedback, and control of process parameters. Within the last few years these efforts have been extended to provide automation, particularly in materials handling and warehousing, in the process as an integral part of it. Although most of these past efforts in automation have been developed internally, and therefore are quite proprietary, there is an increasing tendency to engage outside firms as automation technology becomes more sophisticated. There is a general feeling that, at this time, the technology for materials handling automation after the extruder is quite adequate for present needs and requires only application engineering in most plants to become quite effective. Thus the presently perceived needs for this industry all fall into the general area of monitoring, feedback, and control, even though these are applied today in relatively sophisticated forms.

The important philosophical issue is that it is better to prevent defects in as absolute a sense as possible rather than to manage defects. If a process can be operated with a very high level of defect prevention, then a statistical appraisal of the product is sufficient to ensure that material going to the customer is defect-free. It is believed that very efficient defect prevention requires efficient managing of the polymerization and extrusion steps themselves. This management takes the form of sophisticated monitoring, feedback, and control of temperatures, pressures, and flow rates in all parts of the process above the spinnerette, and of seeking ways to monitor and flag potential process defects (such as pack pressure build-up or deterioration of spinnerette holes) in such a way that changes can be made before off-quality fiber is produced. The usefulness of threadline monitoring on a continuous basis is not at all agreed upon. It is argued, for instance, that continual monitoring of each threadline is useful when a process is extremely variable, but in modern fiber extrusion processes, variability is so low that continuous threadline monitoring hardly justifies its cost. Rather, if defects can be reduced to a very low rate, lower than the current standard of a few defects per thousand pounds of fiber, then statistically determined examination of packages after take-up can catch any process variation that may occur. Continuous threadline monitoring is used for certain high-value products, i.e., medical hollow-fiber membranes and monofilaments and some industrial monofilaments[7].

The general aims of the desired improvements are to prevent process drift. If process drift can be prevented and if raw material variability is at a suitable minimum, then control of the other process inputs such as temperature, pressure, air flow, flow rates and others, is sufficient to guarantee the uniformity of the process. This is a concept that can apply either to filament fiber or staple fiber production. In fact, it is no longer assumed that staple fiber can have a somewhat lower product uniformity than filament.

Current Status: The Textile Industry

As a whole the textile industry has made major progress in process modernization and automation over the last 5 to 10 years. The focus of these programs has been in one of several directions in different companies. Often new technology has been utilized to reduce the number of process steps, but just as often to reduce the labor content of the process. This is particularly evident in the reduction of low skill-high drudgery jobs, but it also applies to the reduction of human intrusion into textile processes, which essentially is a quality argument. Examples of these process improvements include the automation of opening rooms, the installation of chute-feeds and high production cards, the automation, at least in part, of drawing, the introduction of open end spinning, the increasing use of shuttleless looms, the use of automatic systems for handling waste, and the nearly universal use of microprocessor-controlled monitoring and reporting of production variables.

Activities in applying presently available technology to textile processes are being pursued in all of the major process categories.

Opening and Carding

There is potential for the automatic analysis of incoming bales of fiber, particularly if a robot or other automatic device could be involved in the analysis and then direct the bale to the appropriate storage area.

Automated bale storage is a labor saving technology that

leads to less handling as well as presenting the opportunity of coding each bale for future identification. This is important because of the desirability of tracking the identity of the material in each textile process from the initial bale through all of the process steps. This in turn allows one to know the accurate history of all material and all processes and allows the solution of quality control problems on a more rational basis.

Cards can be loaded and unloaded automatically and their output can be conveyed automatically to draw frames. However the need to mix the inputs to draw frames in such a way as to insure good blending is important; these patterns also need to be capable of being changed from time to time.

Yarn Formation

The issues in yarn formation generally point toward reducing the number of steps wherever possible. One view is that devising processes that can successfully do this is more important than devising automated materials handling systems to move materials between currently popular process steps. In general the ideal yarn manufacture process would take fiber from a bale, convert it to a sliver and move it to a spinning process, for instance open end spinning, with automatic transfer of the yarn output either to a warper or to a loom. Emphasis also must be placed on computer monitoring both of quality and of production rate, with the quality monitoring being tied to appropriate feedback and control mechanisms throughout the several processes. The monitoring technology is already available for drawing and the technology for monitoring either open-end yarn or other kinds of yarn is on the horizon.

Another area of considerable importance with respect to monitoring is to be able to determine the need for machine maintenance by continuous monitoring of yarn from, for instance, open-end rotors. This requires that all ends be monitored continuously and that faulty positions be identified immediately. The repair, when it is needed, could be made automatically or, alternatively, that end could be stopped and machine maintenance ordered. A useful system would also monitor and record for long-term gradual deterioration as opposed to short-term problems. In this whole system, the philosophy is to prevent defective material at its source.

In general it is important to measure and to control quality on the spot at every step of the yarn formation process. The computer technology is already available; the real problem seems to be the development of appropriate sensing elements.

There are a number of opportunities in present yarn processes for automated materials handling, including the use of robots. This is particularly true for systems that have reduced the number of process steps, i.e., open end or other systems, but it could also be used with ring spinning technology in which the automation of roving frames is a current need. Connecting winders to large spinning frames is a good idea, and it was demonstrated at the 1983 ITMA, but winders need to be designed to accommodate some flexibility of yarn count; this perhaps could be accomplished by designing winders with space for extra positions. This linking technique allows less handling of yarn packages, which is a distinct advantage, and it also allows better package identification and control.

Although techniques for the continuous monitoring of various stages of yarn processes are available, knowing where to direct that information or what corrective actions to take still is unknown to a large extent. Since it is necessary to be able to identify abnormal parts of the process, the real issue is to determine where in a process faults occur, what their relations with other process steps will be, and the feedback loops in which to exercise control.

Fabric Formation

The issues in fabric formation are almost all in the areas of monitoring and control, inspection, and material handling. Modern weaving and knitting machinery allows the choice of many options and a conscious decision of whether to emphasize productivity or flexibility. The important trend, outside of the design of fabric formation machinery, is the elimination of menial jobs that mostly are associated with materials handling. As an example, fabric handling with driverless tractors is well-proven now, although this is an area in which potential savings other than labor savings are difficult to identify; therefore this may remain fairly standard technology for the future. It also would be desirable to devise systems for automatic doffing of cloth and perhaps automatic cutting, in conjunction with automatic inspection and grading.

Inspection of fabric is thought by many to be a process which should be capable of elimination, if suitable process monitoring and control systems can be invented. There is, however, a difference in philosophy depending on whether a greige mill or a dyeing and finishing plant is involved. In the former case it is probably more likely that inspection could be eliminated if it were to be carried out later on the finished fabric. Probably there will always be a need for some inspection, but it should be automated. In particular it should be capable of viewing the fabric with both the plant specifications and the customer's specifications in mind, changing the latter as the customer changes and changing the former as the fabric style changes from lot to lot. These issues become more important as the textile industry as a whole moves to higher quality and output levels because then inspection speeds become the limiting process speeds. Computer-controlled inspection systems also ought to interface with computer-controlled cutting systems to optimize the cutting of first quality yardage.

Monitoring in weave rooms particularly, but also in other fabric formation areas, is done today almost solely for the purpose of generating management information. Although this is an important function of such systems, what is needed, in addition, is diagnostic monitoring that will locate and diagnose loom malfunctions, preferably before they lead to producing off-quality material. Such systems would be part of a larger monitoring system that deals with the flow of raw material into the fabric formation process as well as with fabric formation itself. The development of truly reliable monitoring and control systems is closely related to the development of automated inspection systems, since the latter depend on the assurance that defects are minimized in the process.

Dyeing and Finishing

In this area there is a general view that the number of discrete processes in dyeing and finishing needs to be reduced drastically, and as many as possible should be combined. The goal would be to make dyeing and finishing a truly continuous process rather than a series of batch processes, each with its own control and materials handling problems. What this implies is the development of some rather sophisticated monitoring and control systems for dyeing and finishing processes generally. Among these systems, the more important seem to be the ability to monitor and control the color of wet fabric, as well as to monitor and control moisture content. The aim is to be able to predict accurately the color of the final dry fabric by measuring its characteristics at the moment it is being dyed. Continuous dye ranges now have considerable automation but are hampered by their inability to run very small lots efficiently. Systems need to be developed that will allow rapid changeover from lot to lot on a continuous range system with a minimum fabric band between the changes. In addition to suitable monitoring and control functions, ways to rapidly alter dye baths in order to change color quickly must be developed. The aim is to produce systems with very rapid response times and this will require precision instrumentation for adding chemicals and controlling the parameters of the process.

For similar reasons it is extremely important to have absolutely uniform desizing and bleaching. In this area there may well be applications also for computer-based monitoring, feedback, and control systems. There also are general needs for reducing the energy cost in dyeing by reducing either the setting or the drying requirements. And for some products, it would be very useful to have dyeing the last of all finishing steps in order to improve order and warehouse versatility.

Materials Handling

From raw materials to the finished product there exist major opportunities for process automation in materials handling. Current applications of robotics are nearly all in this area, for the reason that it seems relatively easy either to identify labor savings or to accomplish a reduction in menial and sometimes hard to fill jobs associated with these processes. An important concept that accompanies automated materials handling is automated identification of the material which, if carried through in all parts of the process, allows the identification of quality control problems with respect to their material source. Much automated materials handling is, or will be, built into the process technology itself; but there remain materials handling problems that will require special technologies, the newest of which are driverless vehicles and robots. On the other hand, it can be considered that automated materials handling is really a second choice to the integration of processes; to the extent that some processes cannot be integrated, materials handling will remain a necessary technology. There is much technology applications development to be accomplished, particularly in the robotics area, which is not yet, in the opinion of some, suited for most textile applications. Particular issues include combining carding with breaker drawing with no intermediate materials handling, the doffing of ring spinning frames, the doffing of cloth at the loom, and the creeling of packages for dyeing, finishing, and warping.

Status Report - The Apparel Industry

It is easiest to approach this topic by starting at the front end of an apparel manufacturing process, so the first topic should be computer-aided design. Unfortunately, in its apparel design context, this is a technology that is not quite ready for commercialization. It has the same goal as computer-aided fabric design systems, namely, allowing the designer to quickly and easily test his or her ideas at a computer console with an appropriate viewing mechanism to see the result of those ideas. The problem in apparel design is essentially that of converting easily and reliably between two-dimensional and three-dimensional information. Technically it would seem easier to take the two-dimensional information from a set of pieces (for instances, for a shirt) and convert them to a three-dimensional image of the shirt's final appearance. However, designers often operate in the reverse mode, that is, they make an illustration of a garment the way they visualize it, and then the problem is to convert that picture to a series of two-dimensional pieces to outline the actual manufacturing process. Research progresses and a number of theoretical reports are available in this area but, as far as commercialization of this technology goes, progress remains slow.

However, the situation is considerably better in the area of computer-aided marker-making, grading, and cut-order planning. There are several systems on the market that allow an experienced operator to take a standard pattern piece, generate other pieces from it on a graded size basis (i.e., grading) and then to pack all of the necessary pieces on a piece of fabric so as to give maximum fabric utilization. Maximizing fabric utilization is probably one of the greatest driving forces for this technology. It is estimated that, for plain fabric, fabric wastage can be reduced to about 3% with this technology compared with 5-6% when human beings do the same job unaided. On the other hand, if instead of plain fabrics, one uses patterned fabrics in his process, then fabric wastage can be as high as 25%, and there is clearly room for improvement here. Although there is no current utilization of this idea, it seems logical that computer-aided pattern packing ought to be combined with fabric inspection information, performed at the textile manufacturer's plant. Such a combination would allow the apparel maker to cut around any defects that might be present and guarantee that those pieces that were cut were defect free; this would effectively increase fabric utilization.

Technologies for cutting fabric are increasingly being debated on a more philosophical basis. The issues are whether, in fact, the traditional bundle system of fabric cutting and material transport can be retained if other parts of the apparel manufacturing system are to be highly automated. The bundle system uses a spreading machine (which some say has not seen any major improvement in many years) to lay and then cut as many as 100 layers of fabric at once. Thus, this system produces 100 layers of shirt fronts, 100 layers of collars, and

100 layers of pockets and these all are bundled together to provide material at each sewing station. The advantage of the system is that it is, or at least is thought to be, a very effective and cost efficient way to cut fabric; therefore, it has inherently low costs associated with it. This assumes that the person who does the cutting, if it is done manually, doesn't make a mistake. One mistake, of course, can ruin 100 shirt fronts. More importantly perhaps, from an industrial engineering point of view, is that the bundle system inherently provides a buffer for each of the other steps in the process. If a sewing machine breaks down or if an operator is sick one day, the manufacturing process can still proceed because of the inherent build up of material available at all stations. The major disadvantage of the bundle system is the long in-process time of the item being assembled. It is said that for some articles of clothing the in-plant time is as high as 20-30 days. This represents an enormous inventory cost and is considered by some to be a major detriment to productivity enhancement. The bundle system also assumes that fabric is identical from one layer to another because the identity of different pieces effectively is lost in the system. To the extent that fabric is not uniform, then there are inevitable wastages because fabric pieces cannot be appropriately matched.

The alternate system, used in some parts of the apparel industry where fabric is extremely expensive and a cutting mistake cannot be risked, is, for the sake of a better term, the one-layer system. As the name implies, each piece is cut individually in this system and then the problem is to route the pieces in the correct order to their assembly points. The advantage of this system, other than the fact that the penalty for cutting errors is minimized, is that this process is actually more amenable to automation and control of the rest of the assembly process. In addition, such a cutting system allows an enormous reduction of in-process time from 20-30 days to, in some cases, perhaps a few hours. It has obvious advantages. The major disadvantage of the one-layer system is that there is no inherent buffer capability. If a piece is missing or if an assembly station is inoperative, then the whole production process is held up. On the other hand, it may be possible to attack this problem by considering the "just in time" manufacturing philosophy that automobile makers use, in which a forward part of the manufacturing process calls to it material just as it is needed and no sooner. There is no buffer concept in that philosophy, but it requires adequate material handling systems and reliable parts interchangeability.

There are major innovations, some of them now quite old, in the area of cutting fabric. Computer-controlled cutting, as exemplified by the Gerber cutter, is a technology that has been around for perhaps 10 years. It is an expensive machine and is utilized only by large companies that can justify having large cutting departments. In its present form, it takes information directly from the computer-aided marker-maker and cuts pieces from layers of fabric in a very rapid and reproducible manner. It also uses a reciprocating knife in most instances, as this has, up to now, been the standard cutting technique. There are more novel cutting techniques available, each of which has been downgraded in the past because of the disadvantages they have in working with the bundle system. Laser cutting is a technology that is used to cut out men's suits, for instance. It can only work really well on one or a few layers of fabric at a time, and it works best on non-thermoplastic fabrics that otherwise would melt at the edges of the cut. (The problem in working with many layers of fabric is that the cut edges tend to fuse together during the cutting process and then they are very difficult to separate.) Water-jet cutting is a technology used in many other fields, such as the shoe industry, the automobile industry, and in a variety of industries where complex shapes have to be cut rather rapidly. Water jets don't work well with the bundle system either, because the power in the jet is severely retarded after the first few layers of fabric. Nevertheless, if the one-layer system were to be adopted more uniformly, then either laser cutting or water-jet cutting systems would make considerable sense. Particularly this is true because either of these technologies can be combined with robotic guidance of the cutting head.

Joining techniques will continue, most people think, to emphasize sewn seams. Consumers expect certain aesthetic and mechanical attributes of sewn seams, and these are difficult to achieve with any other technology. Modern sewing machines operate mechanically much like the one that Howe invented, but they have advanced to the point where microprocessor control is an inherent part of many of them. Microprocessors can control both the stitch pattern and the feeding and ejecting of material once it has been sewn. An emerging technology, not commercial yet, is three-dimensional sewing, in which a sewing head on a robot arm effectively operates in three dimensions on pieces of fabrics that have been placed on a form. There is a considerable amount of research going on in this area. In the area of nonsewn techniques, both adhesives and thermal bonding techniques are important. Again, there seem to be applications for robot systems in this area, although the number of seams that can utilize these technologies probably is limited.

As in the textile industry, materials handling in the apparel industry is a major problem, which can be divided into several aspects. The first is materials presentation at the sewing machine itself. Some apparel companies ascribe as much as 80% of their labor cost to the sewing part of their operations but closer inspection reveals that actual sewing only contributes perhaps 20% of this cost; the rest is materials handling. There are a number of devices already available to bring material in a more orderly fashion to the sewing machine operator. Two such are made by Gerber and by Eaton. In each case, a rather complex overhead track system is loaded with pieces and partly assembled systems and moved by operator demand to the place where sewing is to take place. These systems seem to be logical components of more automatic systems for the future. Many have investigated robots or robot-like systems as devices for materials handling, but the main drawback is the classic limp fabric problem. It seems that it is nearly impossible to design a machine to pick up a piece of fabric, which may have only one of a variety of weights and stiffnesses, and to operate on it in a reliable and consistent manner. There exist devices which can handle certain classes of fabrics. The Clupicker is certainly one of these, and there now exist robotic end effectors that are claimed to be able to operate with a wide variety of fabric types. One of these is included in the Draper project technology. Nevertheless, until the problems of reliable materials flow, not only at sewing stations but between sewing stations, is solved, automation in any real sense in apparel manufacturing will be severely handicapped.

The manufacturing loop needs to be closed with sufficient management information, not only to track a process but also, at least in the ideal case, to determine when a process is approaching trouble. There are a number of management information packages being sold into the apparel industry now which depend for the most part on operator input. Considering that a sewing machine operator has a number of other things to do, one has to wonder just what the quality of the input data to these systems really is. Nevertheless, this represents a considerable improvement from the traditional information gathering technology that has been prevalent.

Directions for Future Technology

The general themes of all of these comments for all three industries are the increasing use of computer technologies, the need for the invention of suitable monitoring, feedback, and control loops and the increasing use of automation in materials handling. These trends arise because processes of the future in these industries will most desirably have minimum numbers of intermediate steps and minimum numbers of people involved in these steps. Another general aim is to design processes with less dependence of unit cost on volume, which therefore would allow management more choice of flexibility or mass throughput for any particular product at any particular time.

In the fiber industry the emphasis will be on developing high-speed extrusion processes that can be fully automatic. Very large packages will be produced of at least 100 pounds, which will mean less inspection. These packages will be loaded directly onto a creel either for warping or directly for weaving or knitting. Boxes and other temporary storage devices will no longer be used. In staple fiber production the fiber instead of being processed into bales, may well be packed directly into a truck and on delivery, transferred to card chute feeds.

In the textile industry, there are those who predict the general automation of processes, particularly materials transfer among carding, drawing, spinning, and fabric formation processes. Particular technology developments may include new spinning technologies that would either replace or refine current open-end technology; these will become necessary because cotton production is producing increasingly trashier material.

Several respondents suggest that replacing the card is a necessary feature of future technologies, largely because cards don't allow for easy material transfer to the next step. An ideal system would send fiber directly to open-end spinning. If cards are not replaced, provision for loading and unloading them automatically will need to be made and, in conventional systems, sliver cans will be transferred automatically to draw frames. The output from these can be transferred automatically to open end spinning frames and robots may become increasingly useful at this stage. In fabric production, an ideal situation will allow changing styles on looms much more rapidly than at present, and it will also involve the rapid transfer of computer-generated fabric designs to looms for trial runs. Fabric grading will be largely automatic with the ability to distinguish not only kinds of defects, but also their placement within the fabric and to characterise those conditions with reference to either plant specifications or customer specifications. Such systems will work on any style and any color of fabric. Dyeing and finishing, but particularly dyeing techniques, will be totally automated, capable of making nearly instantaneous changes from one shade to another and capable of reproducing shades absolutely. The implications for computer monitoring, feedback, and control of these systems are enormous.

Management information will be much more complete and easy to use than it is today. The information gathering used for managerial decision making will be part of the same process monitoring and control information gathering that will be necessary to run an error-free process. There also will be a real need to be able to estimate cost, labor and material savings, and management time in the process of determining whether a new device or technology is worthwhile at a particular cost. The emphasis, in general, in the textile industry as well as in the fiber industry will be on upstream quality improvement, which is more likely to produce a higher quality product than automated inspection systems.

In the apparel industry, emphasis will be placed on more efficient materials utilization, and on speeding up material flow through the factory. This may be more important in the long term than direct labor savings. However, the cost of inventory is so high that any technology that allowed a garment to pass through the system in a few hours as opposed to many days would be very cost-effective. There will be increasing emphasis on automation in sewing and other joining technologies as well as material transfer from station to station. This is the thrust of both of the widely publicized research programs in the automation of apparel manufacturing: the Draper Laboratories program and the Japanese MITI program.

Both of these programs utilize the concept of the "manufacturing cell", in which robots, hard automation, computer-aided design, and computer-driven information flow combine to define a self-contained manufacturing unit. The first is an American project sponsored by the Tailored Clothing Technology Corporation, abbreviated $(TC)^2$, but loosely called the "Draper" project after the Draper Laboratories where the work actually is being carried out [8,9]. The philosophy of this project is that a manufacturing cell exists wherever a sewn seam must be accomplished, and that the important problems are getting material into the system, performing some orientation and/or parts identification, perhaps folding the material, sewing it in a reliable fashion, and ejecting the material from the system. The Draper prototype machine that was completed some time last year (reputed in its present form to cost around $50,000) consists of a number of novel devices to do these things. The important technology seems to many to be in the use of vision systems to identify and orient fabric parts, and in the claimed invention of a robot that can, in fact, reliably pick up single layers of fabric (although in this case it is for folding; i.e., moving one layer over another) and in a sewing head that travels to the place where the seam is to be made. Each of these operations is microprocessor controlled and, at this time, the process has been designed to do only one thing, to sew the sleeves of men's coats. The Draper project investigators claim that the technology is adaptable to a wide variety of apparel assembly processes and certain of its aspects definitely are, particularly the vision system and the materials handling technology. Some in the apparel industry criticize the project, partly on the basis that it represents a very expensive, very specialized machine, and partly on the basis that nothing has been done to change the sewing technology at all.

A project receiving almost equal publicity in this country is a Japanese project in which a group of Japanese companies under the auspices of the Japanese government (MITI) have banded together to form an organization called Technology Research Association of Automated Sewing Systems, which has published a very elaborate program for the complete automation of apparel manufacture--what really amounts to a completely automated factory [10]. Whereas the Draper project has cost on the order of $5 million so far, the Japanese project is reputed eventually to cost $50 or $60 million, and there are many who believe that it will cost much more than this before it is finished. At this stage, planning and some fundamental work are being done in the areas of sewing preparation, sewing and assembly, fabric handling, and production control. Thus, all of the apparel manufacturing process is being attacked, with the desired end result being an automated factory composed of a number of individual work cells, into each of which material is fed as appropriate and taken out as appropriate, with the entire process operating very much as an integrated system. While the Draper project emphasis is partly on the reduction of labor content, the Japanese emphasis is almost completely on the elimination of labor content. Many explain this as the difficulty in Japan of obtaining apparel plant labor, and suggest that this really is an attempt to preserve an industry for which a labor force has largely ceased to exist.

The Future

Are we approaching the day when textiles and apparel will be assembled in an automated factory? This is the goal of many manufacturing industries. The conventional definition of the "factory of the future" is one that combines computer-aided design, hard automation, robotics, and computerized information systems to provide a self-contained manufacturing cell that has as its input computer generated information and has as its output both computer generated information and a product that is uniformly manufactured to a specific quality standard. Whether or not this all can happen in the textile industries is dependent on other, nontechnical, aspects of these industries. If any of these levels of automation are to take place, they almost certainly will take place according to company type and, as a result, many predict that the overall textile industry will become divided into two basic kinds of companies, those that are more manufacturing-oriented and those that are more marketing-oriented. The marketing-oriented companies will tend to be design-oriented, specialist companies who place foot market response above production efficiency. They will react very quickly to the marketplace, they will manufacture high value-added products, and they will use less of the available technology than they might if they were manufacturing larger amounts of their product. However, some technologies will be important to them, particularly in the area of computer-aided design and in management information systems. They will tend to judge automation by its ability to enhance the value that they consider most important, i.e., flexibility with respect to market demands.

On the other hand, manufacturing-oriented companies will be very large and they will make very large amounts of their product. They will be prime candidates for large scale automation and perhaps even for the automatic factory. At any rate, they will have to have the ability for significant investment of capital and this, as much as anything, will determine whether a company can afford to be manufacturing-oriented or not. The primary thrusts of these companies will be productivity and quality and, in order to achieve these, they will try very hard to lower their overall labor costs (not only their direct labor costs) and to lower their in-process time so as to escape the high cost of inventory. They will be very quality-oriented and, even though they are large manufacturers, they will have to retain some of the flexibility necessary to accommodate market demands on at least a longer term basis.

Conclusions

Observers of the textile and apparel industries long ago concluded that some kind of change is inevitable if those industries are to survive. Whether that change is large, as it must be for those companies that intend to be primarily manufacturing-oriented, or rather small, as it may be for those companies that intend to be primarily marketing-oriented, nevertheless, some change is inevitable. Most observers agree that simply controlling imports is a necessary short term goal, but cannot insure the long term survival of the industry. The argument for import control is basically one of giving time for the necessary modernization of the industry. This modernization will lead to a major division of companies into manufacturing-oriented companies and into marketing-oriented companies, and it will be difficult for one company to be both. Although this will lead to a major division, it also may lead to less fragmentation of the industry as, particularly for larger companies, there may be considerable incentive to become even larger through acquisitions and other kinds of mergers.

Finally, the industry will have some major nontechnological needs if this scenario is to be carried out. One is the need for capital. All of these emerging technologies will cost money, and some of them will cost enormous amounts of money. The definition of a reasonable return on investment for production facilities will have to be looked at seriously. It is very doubtful whether many of these technologies can be justified on a 12 or 18 month return, for instance. The ways in which companies obtain capital and, in general, conduct their financial affairs will have to be looked at very critically. The other area that is bound to change is the nature of management personnel. The future that is suggested here requires highly skilled managers but also technically adept managers. Both at the hourly level and at the managerial level, the nature and backgrounds of the people involved in fiber, textile, and apparel manufacturing could change significantly.

REFERENCES

1. Krause, H.W. in A. Heusser, Textile Industries, pg. 110 (Figure 2), September 1983.

2. Hunt, V. D., Industrial Robotics Handbook, Industrial Press, New York: 1983.

3. Bailey, J. R., presentation at Computer-Aided Manufacturing Conference, North Carolina State University, May 1982.

4. Conigliaro, L., Presentation at Vision '85, Detroit, March 1985.

5. Buchanan, D. R., Apparel Industry, pp. 58-60, December 1984.

6. Buchanan, D. R., America's Textiles, pp. 94-98, March 1985.

7. Caban, J. C., Fiber World, pp. 25-32, July 1985.

8. Anon., Textile World, pp. 35-38, February 1985.

9. Purdly, S., Apparel Industry, pp. 40-44, February 1985.

10. Jones, S. H., Apparel Industry, pp. 46-49, February 1985.

Corporate Planning Issues Concerning Robotics in the Textile Industries

John Strand
Amphion, Inc.

How does a company plan for technological change? Can the important issues be identified? In this chapter, the importance of planning is stressed and the significant issues to be considered are discussed.

John Strand has had a long career in the field of technology transfer. From his work in radar technology during World War II to his present position as President of Amphion, Inc., in which he is responsible for the design of automated factories - including a few in the textile industry - he has been involved in the application of new technology over several decades. He received his education at the University of Michigan, where he obtained an M.S. in Electrical Engineering.

Corporate Planning Issues Concerning Robotics in the Textile Industries

There is only one planning issue relevant at the present time. Either plan to automate or wait to die. Our current manufacturing practice with our current bureaucratic management philosophy cannot survive world competition.

It's no secret that our country has a productivity problem. It's no secret that we are in a battle with foreign competition, not only for world markets but for our domestic market as well. It's no secret that our system that once was the best in the world has run its course and is being superseded by a better one. It's no secret that the better system is computer-controlled automation and that only through such automation can we hope to again achieve world supremacy as a producer of industrial products. The only secret is who in America will be the winners and who will let themselves wither on the vine.

I realize that what I am saying may sound self evident. The concepts are simple enough, but often simple concepts are misunderstood and I would like to convince you of what I believe to be a few basic truths concerning the future.

The first is that no one can predict the future precisely. We usually tend to predict that some development that we know about will advance much more rapidly than it does and we usually fail to predict entire new turns that development will take. We can't predict fads but we can detect trends and we can use knowledge of these trends to help us adjust to the changes in our society as they occur.

The second is that we are rapidly changing from an industrial society to an information society. Already there are more information workers in America than there are industrial workers. This trend will continue. This does not mean that there will not be industry. The production of goods will continue because the demand for goods will continue. The reduction of industrial workers will not reduce industrial output any more than the reduction of farm labor reduced agricultural output. Nor does it mean that we will become a service society where we are serving hamburgers to each other. Some authors have equated information handling occupations to service occupations and this apparently has led to some erroneous conclusions about an information society. Information is truly a value adding commodity and much of our future wealth will be in information.

The third truth is that right now we are much better at generating information than we are at using it, so that we have a real danger of drowning in a sea of information while we are dying of thirst for knowledge. The human mind, as wonderful as it is, cannot begin to cope with the information our society is generating without the use of computers to extend our minds. It will be people aided by computers, that will make the automated factory possible.

The fourth truth is that the United States is by far the largest market in the world for consumer goods and will continue to be so for the foreseeable future, at least until the year 2000. Economic forces will dictate how that market will be served.

Those products that continue to require large quantities of human labor will be manufactured where such labor can be obtained at lowest cost. Those products that can be largely produced through automation will be produced near their market to avoid transportation and distribution costs.

Market forces will from time to time be distorted by various vested interests who will persuade their government to circumvent market pressures by protective and restrictive legislation, but these efforts tend to be self-defeating and eventually market forces will prevail.

We have no one to blame for the situation we are in except ourselves. We are largely self-governing and the sooner we learn that laws do not create wealth but only redistribute wealth, the better off we will be. There will always be special interests pleading for protection so they can continue to exist even though they are noncompetitive, and as long as the rest of us support their position we deserve the consequences -- which will be a reduced standard of living.

This is especially true of textiles. Why protect our shirtmaking sweatshops against Hong Kong when Hong Kong can't protect itself against Thailand and Mexico where labor is even cheaper? And how long will it be before the shirtmaking business in Thailand and Mexico falls to the lure of lower wages in Egypt or India? Isn't it better to automate and use our new found technology in the apparel business, such as using computer driven lasers to cut patterns, than to protect obsolescence? I'm not here representing the apparel industry and haven't the faintest notion whether it has the will and capability to survive. I am attempting to speak for textile manufacturers and I sincerely believe that we have what it takes to be dominant worldwide in textiles without government subsidy.

When I said we have no one to blame but ourselves, I mean all of us. If I were to rank us by groups, such as management, labor, and government the management group would be high on the list of the blameworthy. They can blame their near-term bottom line myopia on shareholder demands for good stock performance, they can blame the unions for their failure to establish good working relationships with labor, and they can blame government for laws which do not let the free market operate. These agencies are not without blame but management cannot escape the fact that they have not been aware of the direction that worldwide manufacturing is going and as a consequence they have, for the most part, not taken appropriate protective action. Arthur D. Little's Thomas Gunn was recently quoted as saying, "Senior managers of many U.S. companies present a challenge. They don't understand the manufacturing of ten years ago let alone that of the year 2000, because most came up through the marketing, finance, or legal ranks." He went on to say, "To get where we want to go, however, we'll need to put the control of manufacturing back into the hands of manufacturing people."

It is important that we all understand that we're being swept along by Toffler's <u>Third Wave</u>. It's important to realize that future generations are going to demand more individual freedom, not less. We must realize that energy is no longer cheap. It never really was; we just legislated that it be so. We must realize that we are just part of a worldwide market society that we can no longer control. If we are to survive in this marketplace we must operate in it according to existing market rules. Those who are efficient will prosper and those who are noncompetitive will perish.

Henry Ford invented mass production and performed miracles through the division of labor; where one person might put a washer on a bolt, the next one put the bolt in the hole and the third tightened it. This automization of work is most efficient where one is making millions of identical pieces and must use human labor as the means of assembly. It works best in a world where you can order your Model T any color you want as long as it's black. People really want variety. And our new world, with its high-speed computers, high-speed communications systems and flexible manufacturing, allows this variety to be produced efficiently. The days of mass production are on the wane and probably gone forever as far as U.S. production is concerned, giving way to demassified production where people can order what they want. Division of labor and mass production will still be a valid concept for awhile in third and fourth world countries, but these countries will have to compete with the first world's computerized manufacturing and its automated processes. The lowest cost system will win in the market.

Division of labor can no longer work in the United States. We have finally reached the point where we have complex people doing such simple tasks that they are no longer satisfied with their jobs. This is true because these people have many options through entitlement programs that include not starving while not working.

People frequently voice the concern that robots will take

away people's jobs. When our country turns to automation some jobs will indeed be lost, but they are in fact the robot-like jobs that people no longer want to do. We've all heard about the "blue collar blues". The best way to combat these blues is to eliminate the blue collar. Instead of forcing intelligent human beings to perform jobs that robots can do better, it would be wiser to teach them to perform jobs that utilize their minds.

Only the human mind can perceive exception conditions that the robot programmer failed to foresee and only the human mind can exercise judgment and take appropriate action in the face of unexpected stimuli. Let's get rid of the stultifying jobs and get on with the stimulating ones. All history has shown that such job elimination leads to more aggregate jobs although with major shifts in job type. The new jobs will be more rewarding than the old jobs.

Accomplishing this shift to automation will require a management unlike that found in most U.S. industry today. Today's management was recently described by James Lardner, a manufacturing V.P. of Deere & Co., when he said "Few senior managers are interested in or wish to become involved with technologies that differ from those with which they are already familiar."

In the first place management must be involved with the new technology. Everyone must be involved with the new technology. We are creating a new kind of laborer, one that thinks, and that laborer can only be motivated to perform at high level by intellectual appeal to his or her enlightened self-interest.

In the last analysis, perceived self-interest is the only motivator ever possible. Much labor in the past has been achieved by appealing to the laborers' interest in escaping punishment either here (as in slavery) or hereafter (as in puritanism). Interest in not starving always works but fortunately we have long since passed the point where that is a socially acceptable motivator. That leaves appeal to the mind. We must work together so all participants perceive that it is in their own selfish interests to make the venture successful. Management no longer may treat employees as assets to be utilized most efficiently. An enlightened management will realize that their subordinates will invariably and consistently reflect the behavioral example that is set for them. Unless this example reflects individual worth the system will not work.

The few examples of American firms that take this attitude toward their work forces and their customers offer ample evidence that the system can work. IBM is perhaps the best example of a huge empire that was founded on and still observes a philosophy that has embedded in it the dignity and worth of the individual. Other such companies include Texas Instruments, Hewlett Packard and Tektronix.

If we try to stay competitive by extorting more output from human labor already weary of its condition, we are doomed to failure. The situation is by no means hopeless. The same Deere and Co. executive I quoted earlier went on to say, "I regret I am not just starting my career in what I believe will be the most challenging and most satisfying part of the reindustrialization of America."

This reindustrialization will come about in spite of anything we do. The forces are on the move. Those that move with them will win, the rest will be crushed. Purely financial people have been calling the shots too long. The Japanese have shown us what benefit can be gained by a well orchestrated manufacturing strategy. Their advantage with worldwide marketing has exposed the deficiencies of our system but "we will overcome".

Corporate Planning Issues—Apparel Options

Edwin M. McPherson
North Carolina State University

Apparel companies, as well as some other labor intensive industries, have other options in addition to large capital investment in highly automated manufacturing plants. In this chapter, Mr. McPherson discusses the arguments for and against investment in automation as opposed to establishing plants in low-wage countries.

Mr. Edwin McPherson has had long and varied experience in the apparel manufacturing industry. After his education at University of the South and Vanderbilt University (MA in Economics and Law), he worked for a number of apparel companies including Kurt Salmon Associates, H. D. Lee, Kellwood, J. Schoeman, and RCA, and as a consultant to the industry. At this time he is Visiting Associate Professor in the Department of Textile Management and Technology of North Carolina State University, working with the Apparel Management program in the area of modeling studies of current and proposed apparel processes.

Corporate Planning Issues—Apparel Options

Introduction

Apparel investment choices have broadened substantially over the past ten years. Firms are approaching alternate opportunities in a variety of ways. Each firm's approach is influenced by not only its competition but also by its perception of the future.

Three streams of activity are emerging: 1) investment in low cost labor areas (Far East, 807 countries), 2) higher technology investigations (Europe, U.S., Japan), 3) investment in higher technology (robotics, seaming).

To some extent individual apparel products dictate or control the amount and type of investment made by a firm. Higher style products with relatively short life cycles tend to upgrade the skills of the operator and to keep equipment simple and flexible. Products which have longer lives have more options as to manufacturing investment or sourcing. Competitive activity can influence these decisions. Investment in upgraded equipment, better methods, and training have increased the American worker's productivity over the past decade. However, the same effort overseas is also making imports more competitive.

Automation of sewing activities imposes disciplines upon manufacturing operations which tend to limit product flexibility. One must also consider, when reviewing equipment, that apparel firms often make major expenditures on styling and design - in fact many firms' survival may depend upon "New Looks" which attract the shopper's eye. This activity tends to frustrate combinations of sewing activity.

Labor cost of apparel in this country runs from 20 to 30% of the manufacturing cost. While this is a significant figure, it is not the dominant cost. Manufacturers may find investing in the conservation of materials or the reduction of operating and overhead costs to be sound logically. Marker making and computerized cutting are often justified on cloth conservation. Computerized work flow monitoring seeks to reduce excess costs and some clerical overhead. Both of these investments tend to improve manufacturing flexibility.

Attacking labor costs through investing in capital equipment as a substitute for labor offers the best avenue for competing with low cost labor. Programmable seaming devices sequenced and fed by robots are within the realms of existing technology. The key factors in the use of higher technology are still cost, internal rate of return, flexibility, and alternate business or manufacturing options.

INVESTMENT IN LOW COST LABOR AREAS

Reducing labor costs from an average six dollars an hour to fifteen cents in conjunction with lower cost (or used) equipment can create a significant price edge. As an example:

	U.S. Product		Asian Product
Labor	$ 6.00		$ .45*
Material	8.00		8.00
Overhead	10.00		5.00**
	$ 24.00	30% Frt/Duties	4.04
			$ 17.49
	100%		73%

* Allows for lower productivity

** Allows for accelerated depreciation and some U.S. administrative costs.

In addition to low cost labor, many overseas or 807 locations may offer inducements to relocate, such as tax benefits or low-cost buildings on favorable terms.

Let us pose a hypothetical situation. Company A's business is growing. It has the option of establishing a 500 operator plant using the most modern equipment, including computerized cutting and production reporting or moving into a foreign trade zone and using second hand equipment with a lower level of labor productivity.

	U.S.	OVERSEAS
Land	$ 535,000	99 Year Lease @ $500 per Year
Building	2,000,000	$ 255,000
Equipment	2,750,000 (550 Machines)	1,500,000 (1500 Machines)
Computer Activity	1,000,000	0
Investment	6,285,000	$1,755,000
	100%	27.9%
Estimated Annual Operator Payroll	$6,240,000 (Excludes fringes)	$608,400 (Includes fringes (Sri Lanka)
	100%	9.8%

This example omits taxes, fringes in the U.S., and many other factors. It deals with basic arithmetic. U.S. machines are averaged at $5,000. Overseas machines are averaged at $1,000 (which can be lower in the used market). Building cost in U.S. is computed at $40 a square foot. Overseas building costs (local construction) is averaged at $5.00 a square foot. The U.S. averages, with cutting area, about 100 square feet per operator. Asia averages about 30 square feet per operator plus a cutting area. Offices are left out but should be proportionate in cost.

This cost pattern, with minor deviations, exists in many Caribbean, Asian, and other under-developed areas. Labor may range from 15¢ per hour up to $1.10 per hour. Productivity may range from 30% of U.S. rate to 85% (or in Mexico 100%) of U.S. standards. Start up costs may range from 30% U.S. to 60% U.S., depending upon the support from U.S. Equipment choices can also change. As an example presses can be replaced with electric irons, cloth spreaders with people, etc. When low cost labor is considered in relation to equipment, robots and automation become less practical.

In the above comparison, the capital investment in the U.S. is about double that overseas. The operating capital requirement is about ten times higher for the U.S. than for overseas. At current interest rates, these differences may be more significant than just the labor costs. A recent study by the American Business Conference entitled "The High Cost of Capital-Handicap of the American Industry" indicates that the U.S. has generated an economic environment that focuses managers' attention on short-term payoffs rather than on long-term investments. There are problems associated with investment in many of the low cost labor areas. These range from planning and controlling production from a distance through coordinating movement of material and to maintaining proper quality levels.

As a result, many apparel firms use 807 activity to reduce control problems and improve turnaround time. They may cut garments in the U.S., sew in the Caribbean area, and finish the product in the U.S. Sewing may be done by contractors or firms may establish their own operations. While this approach does tend to improve control and to reduce inventory investment, it usually involves a higher wage base than the Orient.

In any overseas investment there are a wide range of other considerations which are not normally evaluated in the U.S. Items such as political stability, work ethic, legal restrictions, hidden taxes, graft, etc. are just a few of the factors requiring evaluation. Today U.S. apparel quotas also influence plant location.

In summary, location of plants in low cost labor areas tend to reduce the incentive to use automation or robots.

Complex equipment is more difficult to maintain and the probability of training a high tech maintenance staff is lower. Further since some countries, such as Nicaragua, have proven volatile, minimizing the equipment investment makes sense.

Higher Technology Investigations

Apparel research groups or foundations exist in Europe, Asia and Canada. These organizations usually work with the apparel industry, the government, unions, and universities to review and develop technology for use by the sewn products industries. Most of these programs are government or joint government/industry funded. Usually the projects are staffed by a combination of industry, government and academic personnel. Activities range from studies on the penetration of low cost foreign apparel into the country market through funding of research in new technology broadly useful to the industry.

Typical of European activity is Nilas Nilsson's description of Sweden's response to imports as described in a recent Bobbin Magazine article. Twenty years ago Sweden's sewn products industry was impacted by low cost apparel imports. Ten years ago the industry was in disarray with plants closing and jobs disappearing. For the past ten years the government has spent an average of forty million dollars on the Swedish textile and apparel industry. Approximately twenty-six million annually has been spent on social programs to train and rehabilitate workers. (Over half the textile-apparel labor force is now gone.) The balance was spent on taking over a number of textile and apparel manufacturing firms threatened by shut down in order to restructure the industry through forced mergers, or on research and development. Some firms were able to introduce high technology equipment in order to achieve higher levels of productivity.

The Swedish Board for Technical Development and the National Industrial Board in cooperation with the Swedish Textile and Apparel Manufacturer's Association initiated and funded a number of structural and technical development studies. From these studies a number of research and development projects were identified. Research teams at Chalmers University, in cooperation with the Institute for Management of Innovation and Technology, have undertaken the development of a number of automation or semi-automation projects in the apparel industry. Projects range from improvement in sewing techniques through the examination of integrated sewing systems. Government funding has been a key factor in this program.

Similar projects at varying levels of support exist, or have existed, in England, West Germany, France, Italy, Canada and the U.S.A. The U.S.A. studies were sponsored by the Department of Commerce and the National Academy of Science. Georgia Tech, Philadelphia College of Textiles and Fashion Institute of Technology, together with consultants, were used to examine the sewn products problems, as well as to prepare suggestions as to what might be usefully done. Summarized, the U.S. reports recommended investment in research and development in a number of areas.

Other studies, in Europe, ranged from proposing projects within specific apparel segments through generalized approaches to automation in the apparel industry. Universities, governments, and consultants have repeatedly examined the state of the art, seeking points of possible breakthrough to higher levels of technology. Specialized projects, government sponsored, are currently in progress in a number of sewing areas.

Research and Development in Japan is currently directed

toward the development of a complete sewing assembly system utilizing robotics to move and position "stiffened" cloth. In this approach, research appears to be directed toward generating hard goods characteristics in cloth so that material may be moved by assembly line techniques. Again this is a government/industry sponsored project funded by the government.

Industry, both directly and in conjunction with others, does some R & D directed toward particular problems. Usually these investigations are specific and are designed to achieve an immediate return on investment. It is seldom that an apparel firm will do speculative R & D. This is one reason several governments have established research programs.

Investment In Higher Technology

Some apparel studies show that many firms do not take advantage of technology currently available. A Shirley Institute (England) study covering a representative sample of single-needle lock stitch stations determined that approximately 80% of these machines were not equipped with needle positioners and thread trimmers. Equipping these machines with these devices would improve productivity 7.2%. This relatively minor investment ($874 plain vs. $1,616 equipped in 1983) equates to an annual saving of about $847 on $6.00 per hour labor, or a payout of less than a year.

A 1979 Department of Commerce study indicated that the apparel industry spends about 0.05% of sales on R & D. This equates to a total industry average of 2%. Figures of this sort are somewhat misleading. Larger firms do modernize equipment and do spend monies on R & D. Smaller firms usually are so marginal that they cannot increase their capital investment, particularly at today's interest rates.

Investment in a computerized marking/cutting system can run as much as $750,000. Under the typical apparel industry criteria, this project should "pay back" its initial cost from before-tax savings in two years, or $375,000 a year. This return is equivalent to displacing 32 $6.00 an hour operators a year. The probability of achieving this standard is unlikely for major expenditures. However, the larger firms tend to establish return on investment standards related to the life of the investment, and also related to factors other than pure labor cost reductions.

Currently it is not possible to locate and to buy total assembly systems, except for a few garment areas. These sewn products items, for the most part, are not within the core of apparel assembly activity. This core consists of firms and products which must join multiple seams in different patterns to create new, innovative looks attractive to a wide range of consumers at low cost.

Within the main stream of apparel assembly are some 2,500 firms with 6,000 plants with over fifty employees each (Daily News Record, May 11, 1983), very few of whom are capable of large-scale investment in R & D or in high-priced equipment. Perhaps no more than fifty firms in the entire industry are capable of and willing either to invest in large scale new operating equipment or to pioneer new assembly systems. Generally speaking, the sewn products industries represent a low investment sector of the economy. The firms often are the creation of an entrepreneur who enters business on a shoestring and scrambles to stay ahead of his banker.

Some major U.S. firms are making substantial effort to modernize and automate their sewing operations. Some consulting firms have made studies showing that increasing U.S. productivity through the use of known automatic equipment will make U.S. shirt and sweater producers competitive with Hong Kong, Korea and Taiwan. There is a crossover point, according to these studies, where the rising cost of labor and shipping allows the U.S. producer to compete with the more advanced Asiatic competitors.

Even where productivity can be increased to compete with low (or lower) wage countries, the apparel industry still faces the problem of the cost of equity capital and its associated discount rate. One factor in the 50% productivity rate of return normally associated with the industry is the risk of no return. In a constantly changing market, a firm may miss the prevalent styling criteria which are selling.

A two season miss can lead to complete loss of market share and thus to plant closings. Improved productivity does not help under these conditions. This leads to minimizing investment risk by modernizing practices slowly or gradually rather than by total overhaul.

The present application using robotics, displayed by Rimaldi at recent equipment shows in Cologne and Atlanta, fits the pattern of gradual innovation. Two operations were robotized. Both have wide application in jeans. A reasonable return on investment appears to exist. Applications which use a programmable computer chip to duplicate sewing operations also fall within the pattern of gradual evolution.

Summary and Conclusion

Existing technology offers the apparel industry the opportunity to be more productive. This existing technology probably cannot close the gap between production in low wage areas and in developed countries.

Many governments are looking at new techniques of apparel assembly. Some government funding of apparel automation has occurred both here and abroad. Even if there should be a major equipment or system breakthrough, there is substantial doubt that substituting capital for labor will be enough on its own.

The government and the industry will have to address such areas as tax credits, low cost government loans, product flexibility, and the degree to which the industry can stay ahead of overseas competitors.

Under present circumstances it is most difficult for the apparel industry to justify major investments in the U.S. Less money invested elsewhere appears capable of producing the same product for less. Yet several large firms, such as Cluett Peabody, have announced multi-million dollar investment programs in the U.S. At the same time, two major textile firms, Dan River and Westpoint Pepperell, have announced plans to manufacture apparel in the Carribean under the 807 Tariff Convention.

Even though the labor base in the U.S. textile/apparel complex is and has been shrinking, opportunities for creative, innovative producers of quality merchandise still exist. As noted in the Swedish study, some producers survived because of quality and excellent styling.

Textile and Apparel Technology in the United States

Lukas Michel, Sulzer-Ruti, Inc.
Nicholas Williamson, University of North Carolina-Greensboro
Gordon Berkstresser, North Carolina State University
Henry Kelly, Office of Technology Assessment, Congress of the U.S.

The current state of technology utilization and its effect on corporate profitability is discussed in this chapter, which is a condensation of a report to the Office of Technology Assessment on this subject. It is shown that the textile industries affected can be divided into three groups - one that simultaneously is technologically innovative and profitable, another that is more conservative and less profitable, and a third for which the outlook with respect to technological change and/or profitability is less positive.

Mr. Lukas Michel was a graduate student in the Department of Textile Materials and Management of North Carolina State University at the time this work was done. He now is employed by Sulzer-Ruti in Zurich, Switzerland. Dr. Nicholas Williamson is Associate Professor in the School of Business of the University of North Carolina at Greensboro. His specialities are in the areas of management and human factors.

Dr. Henry Kelly is with the Office of Technology Assessment of the U.S. Congress. His interests include the comparative state of U.S. vs. foreign technology. Gordon A. Berkstresser III is Professor and Department Head in the Department of Textile Management and Technology. After a number of years in industry with Fieldcrest Mills and others in the area of marketing and product management, he received his Ph.D. in business from New York University. His research interests include the management of technological change in this country and in others, especially Japan.

Textile and Apparel Technology in the United States

The state of technology in the United States has been under study by the Office of Technology Assessment, Congress of the United States during the past few years. The textile and apparel technology section of that study was prepared by the authors listed above. The full study contains marketing and industry structure variables as well as technology variables. This paper summarizes the aspects of the study which are germane to the question of automation and robotics in the textile and apparel industries.

An "over the counter" characterization of the U.S. textile and apparel industries as declining is premature. The U.S. textile and apparel industries are still the most productive in the world. They have the potential to remain a major part of the U.S. economy. However, this is only possible with certain structural changes and, partly, by relying on foreign technology and the development of proprietary technologies in a "high tech" direction. Automation and robotics are crucial for certain industry segments.

Employment in the textile and apparel industry will decline, regardless of whether the industry becomes competitive with foreign nations. A high technology industry will employ fewer employees, but these will be more highly educated. The demand for management and technology trained people will increase, but clerical jobs will decrease. The reasons for the decline in employment are: import penetration, higher productivity with new technology, and the lack of appropriate marketing aggressiveness of the U.S. manufacturers. These factors influence the various industry segments in different degrees. In the aggregate, rising imports influence the growth of the industry but not employment. The U.S. textile and apparel industry production growth is still at 2%. Saving jobs is not a useful argument to keep the textile and apparel industry going.

Textile and apparel industry segments appear to have entirely different structure and performance profiles. For some industry segments, prospects are good, for others, prospects are bad. Employment, competitiveness, productivity and profitability are the critical variables.

Prospects are good for weaving industries, nonwoven industries, and the tufting industry. Prospects are average for the throwing, winding, and thread industries and the knitted outerwear industry. Finally, prospects are bad for the circular and warp knit industries, the hosiery and knit underwear industry, and the apparel industry, in general.

The following are the reasons for good prospects of certain industries. The weaving industries invest heavily in new technology which increases productivity and profitability. The technology is mainly foreign development. The nonwovens industry is clearly in the early stages of the industry life cycle with heavy exports and increasing employment, whereas the tufting industry is fragmenting due to negative returns to scale. This means that larger firms do not enjoy higher productivities because of their large size. This has, however, not led to decreases in employment. Exports are extensive.

Industries with average prospects are the knitted outerwear, throwing, winding, and thread industries. The knitted outerwear industry has high positive returns to scale in the productivity sense. This causes consolidation, but with no effect on employment. The throwing, winding and thread industries substitute lower cost capital for high cost labor with increasing automation. There will be further investments in new technology in mainly large scale plants. The industry is competitive. One can expect decreasing employment due to technologies.

Mainly bad prospects face the apparel and knitting-related industries. The hosiery and knitted outerwear industries are disinvesting due to imports and not technology. In the circular and warp knit industries, both technology and imports reduce employment. For example, highly flexible and innovative marketing of sweaters by foreign firms has gained a large share of the U.S. sweater market, but the foreign firms have no technological advantage over U.S. firms. Furthermore, technological advances do not improve the financial performance. The apparel industries are depressed by large imports. New technology reduces the labor content, but does not always improve productivity.

In general, the textile industries can benefit from large investments in new technology which is mainly produced in

Europe and Japan to improve productivity and profitability with a reduction of labor content. The apparel industries can reduce labor content to a certain point. However, this has not, so far, significantly improved productivity. In the aggregate, both industries will face large decreases in labor content. Some apparel industries must benefit from their potential in marketing by getting international marketing expertise. There are potential growth areas in the textile and apparel industry.

Productivity in the U.S. textile industry has increased faster than the industry average over the past twenty years. The potential for further increases in productivity is high, but increased productivity does not necessarily yield short run financial improvements. Long-term planning is needed, and a certain independence from the financial markets is beneficial. Privatization of textile companies is very common. Productivity increases are mainly geared towards reducing the labor content in production. Higher productivity machinery is mainly supplied by overseas manufacturers. The U.S. lost its competitiveness in the development of textile machinery. Further developments include system-wide improvements in materials handling and more efficient work flows rather than just invention of new gadgets.

Productivity increases in the apparel industry are slow. They all emphasize the reduction of labor content. The sewing process is the limitation to higher productivity increases. There have been no revolutionary improvements in systems to join fabrics, including sewing systems. System wide improvements promise higher returns than "futuristic" technologies.

Improved management and control systems can drastically improve productivity of the textile and apparel industry, overall. Inventory control, the optimization and speeding up of order-to-delivery time are crucial. The time from order placement to delivery can be decreased by improved cooperation of the different industry segments. The fragmentation of the industry must be decreased in the sense that firms integrate vertically to achieve economies of scale in productivity and flexibility with quick response to the market. Vertical integration is not necessary in the sense of finance and ownership. However, this implies that hostility among the industry segments disappears. Most system-wide improvements require coordination between the industry parts without which the full impact of these technologies will not be felt.

Productivity in the textile industry is the most dominant factor in the competition among U.S. producers. The overall productivity is somewhat less important in the apparel industry against foreign competitors since there are other factors like exchange rates and high labor costs which dominate. Productivity advantages over competitors can only be reached with constant upgrading of machinery. Since the technologies in these industries, particularly in the weaving and spinning sectors, change rapidly, the machinery becomes obsolete in short time periods. The capital requirements, for example, for approximately five year cycles in the spinning industry, are enormous. Only financially strong industries are able to upgrade their production technologies without assistance.

There are three major technology related factors which keep the U.S. textile and apparel industries from being competitive and dominant in the international markets.

1) The U.S. textile and apparel industries are dependent on foreign suppliers for machinery.

2) The U.S. textile machinery industry is rapidly declining and has lost ground in major machinery markets in

some areas, such as spinning and weaving. There are no substantial American machinery producers in the market in these two areas.

3) The U.S. textile and apparel industries receive the same technologies as do competitors in the world markets. The U.S. textile and apparel industries have no technological advantage over their foreign competitors.

The development of textile technology indicates that productivity in the textile industry is above the average of all U.S. manufacturing industries. Unfortunately, this is not true for earnings. The technologies, however, are not to blame for the lack of performance. For example, loom speeds have almost doubled since 1970. Tighter international markets, imports, and the lack of appropriate marketing are more likely the cause of unsatisfactory financial performance of the U.S. textile and apparel industry.

The use of the newest technologies is not restricted to industrialized nations. Developing nations are rapidly increasing their capacities with the newest technology available. Machinery manufacturers are likely to shift their emphasis to the Far East, away from the U.S., due to higher growth prospects.

Small advantages in productivity which enable firms to be more competitive are the key to an industry which competes on a cost basis. Labor cost differentials still are a high disadvantage for U.S. manufacturers. However, a positive aspect is that in many segments of the textile industry capital intensity is very high and labor costs become less important due to automation.

New technologies in the textile industry reduce labor

content, and improve productivity. In the apparel industry, the labor production component is reduced without substantial productivity increases. The limits to labor reductions in the apparel industry are set by the sewing operation. It is possible to almost fully automate some textile processes. One can expect that the textile industry will be more successful in becoming capital intensive than the apparel industry. The technology is available, and there are considerable developments going on in all parts of the world. In general, apparel technology still lags substantially in automation.

The evolution of technologies is not at an end. Automation and robotics will further increase productivity and reduce labor. During the past decade, however, the industrialized nations, especially the U.S., lost ground in the world markets despite increases in productivity. While capacities were merely unused in developing nations up to now, they increasingly attain full capacity utilization. There is a high potential for increases in the production quantities of developing countries. The U.S. must be aware of this closing gap.

Major issues on behalf of the U.S. textile competitive situation are related to the level of the technologies in the different industry segments. The question is whether U.S. textile mills have technologies on the newest level. In general, the level of U.S. textile technology is considerably lower than in Europe, but both continents have superior technology levels to the developing nations. The gap, however, is rapidly closing. Developing nations, especially Korea, Taiwan, and Hong Kong are spending heavily on the newest technologies at a higher rate than the Western nations. These three Far Eastern countries can no longer be counted as developing nations anymore with respect to their textile and apparel industries.

Dollar wise, the weaving industry is the largest spender on new technologies. These investments are concentrated among the ten biggest corporations. The ten largest companies account for 85% of these expenditures. It appears that the upgrading of machinery is limited to large scale operations.

Besides upgrading textile and apparel machinery, there are three traits in current development that go through all the different manufacturing processes. These are the automation of all processes, monitoring of the complete manufacturing process, and the integration of all manufacturing technologies in flexible units and fewer steps. Productivity increases and reductions in the labor content are more likely possible through further development of the current technologies and system-wide improvements than through expensive "futuristic" technologies such as molding garments to individuals in zero-gravity environments.

Most textile machinery is imported from EEC countries and Japan. Imports of textile machinery are concentrated in the fabric and yarn industries as well as in some other industries. The American market is still the most important market for all machinery producers. However, some Swiss machinery manufacturers realize the limited growth in the U.S. textile industry and are shifting towards developing countries.

West Germany and Switzerland together account for two-fifths of the world exports in textile machinery. Czechoslovakia and Japan have emerged as important competitors. The concentration is rather high in the machinery industry. The Swiss based Sulzer corporation recently acquired its strongest competitor, Ruti. Moreover, Sulzer-Ruti has close ties with Toyoda, the Japanese leader in looms as well as to British air-jet technology.

These impressive market positions have been achieved with high R&D in their respective fields and traditional reliability in service. The American machinery producers basically lack any momentum in this respect. Most of the European success is also due to their service. All the big machinery producers have service facilities in the heart of the U.S. textile industry. The whole staff speaks English. If necessary, spare parts can be flown in within a reasonable amount of time.

The success of the American textile industry depends heavily on these machinery manufacturers. However, this increasing dependency should be a major concern of the American producers. A key to competitive U.S. textile industries is to slow down the diffusion of new technologies. This would mean to emphasize in-house, proprietary technology.

It is reasonable to believe that the textile companies are unable to invest in process R&D. Furthermore, the market positions of the textile machinery producers is such that a natural dependency is inevitable. The apparel industry, however, can benefit from developments in the electronic industries for further applications of computer controls. American resources are enormous, if used.

In general, the U.S. textile and apparel industry is largely dependent on foreign machinery which is sold all over the world. Developing nations will soon have the same technology level as the U.S.

The structure and performance of the different industry segments in the U.S. textile and apparel industry vary widely. New technologies influence the structure and performance of these industries differently due to the variety of the competitive forces. Also, the structure of the industries is likely to change rapidly with the introduction of new technologies.

Past data indicates that there have been significant differences between the textile and apparel industry's profitability. The textile industry's profitability has been considerably below the apparel industry's profitability. A reason for the differences in earnings might be in the higher cyclicality of the apparel industry's earnings. Apparel purchases are more volatile to the business cycles than the primary textile industry. One might expect, therefore, higher fluctuations in apparel earnings. Earning changes in the apparel industry influence the textile and fiber producer's earnings with a time lag but somewhat smoothed out.

On the other hand, there are no differences in the average productivity of labor between the two industries. Contrary to popular belief, the apparel industry is not less productive than the textile industry. This is even more interesting when the different levels of capital and labor intensities are compared. The textile industries are considerably more capital intensive. As a consequence, productivity is not an argument for the observed reductions in the labor force. It is rather the escape from cost disadvantages in the U.S., thus the labor cost differential. Productivity increases come as a "free gift" with further automation and robotics.

Among the textile and apparel industries, one can observe significant differences in the levels of labor intensity, capital intensity and technological volatility. The apparel and textile industry's levels of labor intensity are different, but not as widely spread as the capital intensity levels. Gross value of fixed assets per employee is about five times as large in the textile industry as in the apparel industry. The capital requirements are, therefore, much higher in the textile industry. Moreover, this enhances the industry's entry barriers. The apparel industry's entry barriers are very low, whereas, in the textile industry, there are considerable capital related entry barriers and scale related effects.

Industry concentration in the textile industry is much higher than in the apparel industry. In the textile industry, there are more large firms dominating the market, whereas the apparel industry consists of smaller, but a larger number, of firms.

Interestingly enough, the industry's firm size adjusted concentration, and the scale assessment levels of the two industry groups are not significantly different. When the firm size differences are ruled out, the concentrations of the four largest firms are not different for the two industries. This means, that in both industry groups, there are a few dominant firms on the market. The respective concentration levels in both are high. In both industries, the four largest firms account for 50% of the respective markets.

The relative levels of economies of scale in the two industry groups are similar. It is not the case that in one industry mainly large or small scale operations dominate. However, one can observe a wide variety of economies of scale in both industries.

Regression analysis indicates that in the apparel industry, monopoly profits (excessive profits after reinvestments and transfers to the shareholders) are literally not responsive to changes in the labor intensity. In the textile industry, monopoly profits are negatively related to labor intensity.

These results clearly indicate, that for the apparel industry, the labor intensity is not directly responsible for the profitability performance. Apparel industries that use more labor as their production factor are no better off than apparel industries that are relatively more capital intensive. On the other hand, in the textile industry, the more labor intensive production facilities are clearly worse off. As a consequence, other factors, for example, the increasing competition from imports, are more likely to influence the profitability of the apparel industry. Obviously, apparel industries of different degrees of labor intensity have been equally capable of protecting themselves from competition. Textile industries which are less labor intensive benefit from their technological advantage. This is reflected in the fact that they are more likely to behave in an oligopolistic manner, and thus, with higher concentration. The trend in the textile industry should, therefore, be in increased expenditures in capital to constantly upgrade the technologies.

In general, industry concentration is invariant to profitability. Firms with large market shares face the same struggles in generating enough monopoly profits as small share firms. The tendency, however, is that the apparel industries' profitability more positively responds to industry concentration. Thus, apparel firms with high market shares tend to be more profitable and are more successful in the battle against imports than apparel firms with small market shares.

The amount of economies or diseconomies of scale has no effect on the profitability of the industry. This means that the behavior of the industries with respect to enlarging or downsizing the operations has no effect on the profitability. The argument that larger plants are more competitive than smaller ones does not hold in both industries with respect to profitability.

It is apparent from the industry structure profiles that no textile or apparel industry is identical in the way it can and will exploit new technologies, especially automation and robotics. Not all textile and apparel industries are able to drastically improve their competitive position with new technologies. In these cases, market forces are such that imports are at levels where inevitably small improvements in cost and product differentiation have only marginal impact. This is where critical market shares have already been captured by foreign firms. The most apparent examples are in knitting and in the whole apparel sector. In the relatively capital intensive industries, such as the weaving and spinning industries, automation can and will enhance the competitive position of firms which can afford it. Also, the necessary and feasible technological changes are closer at hand in the latter industries, to be beneficial within a reasonable amount of time.

Winding technologies took big steps in the past years. Nearly full automation is or can be reached soon. The spinning process still lags in full automation. However, immense improvements are under way in the basic spinning technologies. Rotor and air-jet technologies eliminated steps in the production process. The reduction of steps in the textile production process is the most efficient source of cost reductions. New weaving systems drastically speed up the fabric forming process. High segmentation of the textile machinery industry, however, hinders efficient development of joining technologies in the sense of improving the overall textile manufacturing process. Technological developments are merely further refinement of current machinery.

Entirely different ways are pursued by the newly emerging textile industries, the tufting and the nonwoven industries. As in every early stage of the industry life cycle, proprietary technologies are developed in-house. The benefits are apparent in the impressive domestic and international competitive positions of the U.S. firms which are at least partly a result of efficient technologies.

The apparel industries have yet to overcome the technological disadvantage of the needle-sewing process which requires as much labor as sewing stations. Rumors of prospective revolutionary sewing robots tend to neglect the commercial value of such gadgets. Unless these technologies are cheaper than labor, their practical value is not evident. Therefore, materials handling is where efficient technological improvements have economic value. However, this sets a limitation for representative competitiveness in the U.S. apparel industries.

New technologies are likely to change the economies of scale in manufacturing of the respective industries. The weaving industry is the most apparent example. Air-jet weaving technology drastically enhanced economies of scale. However, if this happens in an uncertain market environment where smaller but different production batches are demanded, flexibility might be advantageous. The apparent tradeoff between flexibility and large economies of scale is the subject of substantial research in suitable weaving technologies in Europe and in Japan.

Changes in the industry structure can be expected, once the development of joining technologies is at the point where spinning and winding processes, as an example, can be performed without intermediate steps. One might expect more vertical integration caused by more efficient materials flow between the spinning and the winding process. Other examples are likely to have the same effect as weaving directly from spinning frames. These ideas obviously are yet to come, but they will change the industry structure. Specialists will be less in demand and firms will vertically integrate to coordinate the workflow.

It is also expected that more vertical integration raises entry barriers. If this is not too late, foreign competitors might find it more difficult to enter the U.S. markets. On the other hand larger economies of scale distract from necessary flexibility in the marketing and manufacturing sense. It is important not to be undermined by quicker foreign suppliers. The degree to which automation is introduced must be carefully evaluated with respect to changing economies of scale, flexibility, entry barriers and overall industry attractiveness.

The mature textile and apparel industries operate with technologies which are mainly foreign produced. Weaving and spinning machinery industries are concentrated in Europe and Japan. Diffusion of technologies can hardly be halted, such that first mover effects are crucial for the competition, but their effects are not long lasting. These technologies can not be characterized as mature. For example, the introduction of the Sulzer projectile loom resulted in changes of industry structure. On the other hand, apparel technologies are mature. Technological development is slow and improvements merely concentrate on current technologies.

Automation and robotics benefit the various textile and apparel industries to different degrees. Mature industries merely require heavy automation and robotics to gain cost differentiation. In the growth industries gains in product differentiation and cost advantages can be reached with process innovation and the development of completely new technologies. Changes in technologies are most likely to affect industry structure. The most apparent tradeoff must be made between necessary flexibility and large economies of scale.

There is a compelling need to keep closer to changing

markets. The volatility of the markets is likely to increase. Demographics and higher income lead to higher demand, but the higher demand is more likely in higher quality and variety. This will result in reduction of the demand for "commodity like" products. Proactive product development through thorough understanding of the consumer's needs will dominate. This is where the U.S. producers appear to lag behind many foreign suppliers. Some of the foreign suppliers seem to understand the U.S. market better than the American producers. Certain industry segments prove that there are potential growth and export areas in the textile and apparel industries. Economies of scale in marketing, for example, through the cooperation of the fiber industry's international marketing expertise and the consumer product or industrial product firms, need to be established.

An industry capable of holding international markets may look entirely different from today's in its interest in research and the ability to absorb technical ideas, the sophistication of its management in production and marketing, and the way it uses its labor.

Although it is impossible to predict for the entire industry what the future holds in store, it is possible to predict what will happen to certain industry segments under different environmental conditions.

If the textile and apparel industries receive long-term protection from foreign competition where there is little motivation for change, it appears that these industries will continue to invest in new technologies in order to decrease labor inputs, but there is no evidence that increased research and development will be undertaken to produce the new technologies in the U.S. Further, there will be no motivation to change the "arms-length" or adversary relationships which exist between key elements in the channel such as textile mills and apparel plants. The history of all industrialized nations shows that long-term, open-ended, blanket-type protection from competition places the undue burden of very high prices on consumers and thus more problems than it solves for the nation.

On the other hand, removal of all barriers to the free and open entry of textile and apparel goods to the U.S. market, while continuing restrictions such as minimum wages, safety regulations, and anti-trust regulations on the U.S. producers will not serve the national interests as it would produce a rapid loss of over one-half million jobs and plunge the nation into a serious recession. Further, such action could result in the inability of the domestic industry to meet the needs of the military services in time of national emergency, as has already happened in the footwear industry. According to the L.A. Times - Washington Post News Service (February 10, 1985), "But the cobbler's trade has fallen on hard times in recent years because of proliferating imports, and the GAO said it was doubtful that the industry could meet the military's needs in a crisis." So, "... in the event of a major military mobilization, the U.S. fighting man may have to go to war in his stocking feet."

Since neither complete protection from foreign competition nor the absence of any protection will service the needs of the United States, it is necessary to create an environment in which the highest possible level of competition is preserved so as to best serve the needs of the consumers of the textile and apparel products while preserving those industry segments which have the capacity to become competitive with foreign sources in the long-run as well as those segments which would have strategic importance in time of military crisis.

For segments which appear to have good future prospects,

such as weaving, nonwovens, and tufting (particularly in home furnishing and industrial textiles), the sense of American "fair play" indicates that they should continue to be protected from "dumping", particularly of goods subsidized by foreign governments. The intense competition to which these industry segments are exposed will certainly motivate them to increase technological developments and should move them toward development of global marketing strategies. In spite of the current problems which the high value of the dollar creates, there are many opportunities for export of goods from these segments than are currently pursued.

Although characterization of the U.S. textile and apparel industries as "sunset industries" is premature, the potential of some segments to remain a major contributor to the U.S. economy rests on joint action between the private and public sectors.

REFERENCES

Annual Meeting of the American Association for Textile Technology, May, 1984, New York.

Arpan, J., Toyne, B., "The Global Textile Industry", George Allen & Unwin Ltd. Winchester, MA, 1984.

Bowen, D. A., "A Time for Change for Textile Machinery Makers", Textile Industries, Nov., 1982.

Buchanan, D. R., "Direction of Technology Change in the Fiber, Textile and Apparel Industries", North Carolina State University, Raleigh, 1984; Apparel Industry, pp. 58-60, Dec., 1984; America's Textiles, pp. 94-98, March, 1985.

Census of Manufacturers 1972, U.S. Department of Commerce, Bureau of the Census.

Census of Manufacturers 1977, U.S. Department of Commerce, Bureau of the Census.

Fairchild Textile and Apparel Financial Directory, 9th Edition, 1982.

Fairchild Textile and Apparel Financial Directory, 10th Edition, 1983.

Fitzpatrick, J., "Why the Textile and Clothing Industries are Shifting to the Third World", Long Range Planning, vol. 16, No. 5, pp. 42-55, 1983.

Freeston, W. D., "The Competitive Status of the U.S. Fiber, Textiles, and Apparel Complex", National Academy Press, Washington, D.C., 1983.

Gale, B. T., "Can More Capital Buy Higher Productivity?", Harvard Business Review, July-August, 1980, pp. 78-86.

Gale, B. T., "Productivity Benchmarks", The Pimsletter, No. 21, Strategic Planning Institute, Cambridge, MA, p. 3.

Jackson, Thomas J., "Textile Machinery", U.S. Industrial Outlook 1984.

KSA Perspective for Textile Management, Kurt Salmon Associates, May 1982.

Krause, H. W., "Entwicklungen im Websektor unter spezieller Beruecksichtigung des Wollsektors", Eidgenoessische Technische Hochschule, Zurich, 1984.

Locher, H., "Textile Industry - Quo Vadis", A Presentation at the Textile Research Institute, 53rd Annual Meeting in Charlotte, N.C., May 3, 1983.

Olson, R. P., "The Textile Industry: An Industry Analysis Approach to Operations Management", Lexington Books, Lexington, MA, 1978.

Porter, M. E., "Note on the Structural Analysis of Industries", Intercollegiate Case Clearing House, Boston, MA, 1975, pp. 3-10.

Porter, M. E., "Competitive Strategy", Macmillan Publishing Co., New York, 1980.

Porter, M. E., "Competitive Advantage", Macmillan Publishing Co., New York, 1985.

Schoeffler, S., "The Unprofitability of Modern Technology and What to do About It", The Pimsletter, No. 2, Strategic Planning Institute, Cambridge, MA, 1980.

Seidel, L. E., "Modern Loom Population and Location", Textile Industries, January 1982.

Strolz, H. "Der Textilmaschinenbau im Spannungsfeld Textil-industrieller Strukturverschiebungen - Tendenzen und Perspektiven", ITMF, Zurich, 1984.

Williamson, N.C., Berkstresser, G. A., "Textile Concentration: A Cure for Productivity Ills of the Textile Industry", Working Paper for Office of Technology Assessment Report. 1985.

The Present State of Automated Sewing System Development in Japan, Western Europe, and the U.S.A.

Gordon Berkstresser
North Carolina State University

Kazuo Takeuchi
Tokyo Economics University

A world-wide view of current research efforts toward the development of automated sewing systems is the subject of this chapter. Major development efforts are underway not only in the United States and in Japan, but also in a number of Western European countries.

Professor Kazuo Takeuchi is Professor of Economics and Business at Tokyo Economics University. His research interests include comparing management styles and technology transfer efforts in Japan and in the United States. Dr. Gordon A. Berkstresser III is Professor and Department Head in the Department of Textile Management and Technology. After a number of years in industry with Fieldcrest Mills and others in the area of marketing and product management, he received his Ph.D. in business from New York University. His research interests include the management of technological change in this country and in others, especially Japan.

The Present State of Automated Sewing System Development in Japan, Western Europe, and the U.S.A.

Industrialized countries, particularly Japan, Western Europe, and the U.S.A. are facing the realization that the very existence of their domestic apparel industries is being threatened by imports. Further, large portions of their primary textile and fiber industries that supply the apparel industry are facing difficult times. The high labor content which current apparel manufacturing technology relies upon, coupled with quality and flexibility considerations has made it imperative for industrialized countries to search for new ways of producing apparel if they are to avoid following in the footsteps of nations such as Sweden, which, in a period of 15 short years went from importing 20% of its apparel to importing 80% of its apparel.

The industrialized nations feel that there are very real advantages in maintaining healthy domestic apparel industries.

They include defense needs, employment of labor, creation of tax base, and avoidance of loss of suppliers (fabrics, threads, zippers, buttons, trims, etc.) from the domestic base. For these reasons, Japan, Western Europe, and the U.S.A. are currently engaged in serious programs designed to produce automated sewing systems which will allow domestic producers in the industrialized nations to survive. The Japanese program was initiated by, fully funded ($60 million) by, and continues to be coordinated by the Ministry of International Trade and Industry, a branch of Japan's National Government. The Western European program was initiated by, received 50% funding ($30 million) from, and continues to be coordinated by the Division of New Technologies of the Commission of the European Communities. The other 50% funding comes from participating industrial firms. The U.S.A. has no formal program, but several groups are currently engaged in research and development activities. The Department of Defense and the Department of Commerce have each supplied some research funds, as have state governments, (funding from all public sources is estimated to be less than $15 million), but the majority of the funding is coming from private rather than public sources, and there is no formal coordination or leadership for this effort in the United States.

The history of the Japanese program has been thoroughly documented in numerous articles in Bobbin magazine. Certainly it is the most coordinated, springing as it has from a collective society where integrated "families" of firms cooperating with the federal government are a long tradition. The Japanese led the way in pointing out the need for such a large research and development effort and they have made a very real contribution to the rest of the industrialized nations in the free world by forcing them to respond to the challenge of the Japanese effort, which continues to make progress, although not at the pace originally predicted.

In Tokyo, Japan, on April 23, 1985, conversations with Tokyo Juki President Takeo Yamaoka centered on apparel automation. We agreed to see him in Cologne in June to look at the "3-D" sewing head which he said would be unveiled there. This is a very light (about 3 lbs.), flexible sewing head that can be moved easily. Mr. Hiroshi Nakamura, Assistant Manager, Planning and Marketing Department (who had just returned to Juki from a stint as Technical Division Manager of the MITI Automated Apparel Project which he helped set up), said that Juki was still a "couple of years" away from making the sewing needle self-threading. Mr. Yamaoka said that Juki could develop each of the individual segments of technology outlined in the Japan MITI project within three or four years, but that getting the total system to work together would take much longer and cost 1,500,000,000 yen (or $6,000,000 U.S.), and that only a few (three to five) firms could afford it. Mr. Yamaoka felt that Japan, the U.S., and Europe should cooperate to develop automated apparel technology for the benefit of all free-world industrialized nations. He felt that the dissemination of technological information on the global scale was so rapid that trying to restrict new developments to domestic markets was unrealistic. Mr. Yamaoka said that every November, all the member companies of the MITI project meet in Tokyo and make full disclosure of all activities for the prior year, but that disclosure is made only once a year, at that meeting.

On April 24, conversations with Yoshimi Ohmori, Manager, Technical Department of Technology Research Association of Automated Sewing System disclosed many important facts and developments. First, Mr. Ohmori is "on loan" from Juki, just as his predecessor, Mr. Nakamura was, so this shows the influence of Juki in this project.

Second, Mr. Ohmori answered the question about dissemination of information by telling us that although Mr. Yamaoka's statement about public disclosure (in November only) of the previous year's progress was correct, there is a private meeting of the members of the project each July at which time decisions are made regarding what will be made public in November. We later learned that Juki President Yamaoka would become President of the MITI project in May of this year. This is entirely consistent with the fact that Juki has a great stake in developing new technologies in 3-D sewing and high-performance pressing.

The most important information gained from this conversation is that part of the program has been delayed for one year. Because of technological difficulties and tight budgets, the system assessment, scheduled for 1986, will be delayed until 1987, allowing an additional year for technology development.

The funds for the project flow through two government agencies and there has been a problem with the flow of funds through the small enterprise administration portion of this process, which has, according to Mr. Ohmori, delayed the technological progress in sewing preparation (the first of the four so-called "Elemental Technologies" to be developed). It is difficult to understand why the assessment of the system should be delayed by one year. If the problems are budgetary, would not an evaluation which moved the bureaucrats off "dead-center" be valuable? After all, MITI did appropriate $60,000,000, and if the bureaucrats who administer the funds are delaying the entire program for a year because they are inefficient, would it not be wise to call attention to this? If the problems center on the inability to solve some technical problems in evaluating and stabilizing fabrics in the preparation phase, would it not be wise to publicize this in an attempt to focus attention on the problem in hopes that some bright person or persons might have an idea which would help solve the problem? In fact, would not the only rational reason for postponing assessment for a year be to defer disclosure of technological progress until patent protection is obtained; or if patent protection seems questionable, to be in process of being able to immediately supply machines and thus gain market share pre-eminence by being first to introduce and deliver new technologies?

In addition to the research in the "Elemental Technologies", there is an item of research called the "Total System", and Mr. Ohmori said that this segment was not being delayed because slow funding does not affect it as badly as it does the "Elemental Technologies". In "Total Systems" research, studies have been made regarding costs of locating plants near big city markets versus rural locations and the effect of a total automated system on labor, management, and their relations. Some material on these items is expected to be published this year.

The fourth "Elemental Technology", termed "Production Control Technology" implies that information flows from the retailer back through the system, but close questioning of Mr. Ohmori revealed that, although the Japanese considered trying to study the entire system, they felt it would be a difficult task, that data would be hard to get, and so they are unwilling to put forth the effort necessary to develop a mathematical model using computer simulation to assess a "quick-response" system. Because the Japanese system has many more levels in the marketing channels, it would be far more difficult to get accurate data from retailers and trace it back through the channels to the apparel manufacturer and fabric producer. According to Japan Apparel Industry Council's Mr. Suganuma, there is no such data in Japan and there are no plans to attempt to collect it. Thus the Japanese have taken the position, in their project, of taking the raw material as a "given" and start with inspection for defects and work forward to determine how fast garments can be produced.

Other Japanese who have said that they feel that the U.S.A., Japan, and Western Europe should cooperate in automated sewing systems research include Mr. Seiichi Iki of the International Affairs Department of the Japan Chemical Fibers Association and Mr. Ichizo Kitano, Executive Director of the Japan Export Clothing Manufacturing Association.

The Division of New Technologies of the Commission of the European Communities was in process of evaluating proposals for automated sewing systems research to fund the first $10 million (U.S.) when Berkstresser was in Brussels June 3 and 4, 1985. Mr. Cyril L. Silver, Head of the Division of Development and Application of New Technologies said that the EC's $10 million had to be matched by $10 million from industrial participants. Next year about $14 million will be allocated by the EC and will also have to be matched.

Mr. Silver said that he felt that Europe was far behind Japan and the U.S. but said "Come back next year and I'll be able to show you something." Silver feels that medium-sized firms will not benefit and will, in fact, be hurt the most by automation in apparel. The small firms are flexible and have learned the art of survival in changing market conditions, the large firms will find the capital to invest, but the medium-sized firms have neither the training in flexibility nor the access to capital.

At the Cologne International Apparel Machinery Exhibition, June 4-6, the European manufacturers were showing sophisticated equipment for specific tasks which was ready for delivery but the Japanese were showing futuristic technology concepts which are not yet ready to put into production lines but which tend to make apparel manufacturers defer purchase of all but immediate-need replacement equipment until they see the end results of the Japanese development efforts. For example, Tokyo Juki President Takeo Yamaoha, who had invited us to see him in Cologne, showed their new "3-D" sewing head, as promised. But it was not sewing anything. Fabric was pinned to a form with exposed areas for seams and the "3-D" sewing head was tracking along the seams with the needle operating but it was not threaded.

It is too early to predict how the European project will progress, but the decision of Sweden's Chalmers Institute to discontinue work on a proposal to the EC program because of commitments to proprietary research, indicates there is a high probability that national self-interests will be a problem.

In the U.S.A., technology development projects such as Textile/Clothing Technology Corporation's development projects focus on fairly rapid commercialization of technological developments such as the automatic sleeve maker. This approach, particularly since there is no strong leadership/coordination role in the U.S.A. runs the high risk of redundancy of research in an effort to "score" and patent new developments.

In the U.S.A. the channels of distribution are much shorter than those in Japan. So while the Japanese are not prepared to build models of the entire system of distribution, the availability of data in the U.S.A. may permit such an endeavor. Dr. Manual Gaetan of Bobbin (among others) has proposed that such an effort be made. The benefits of a model to help solve critical path problems regarding costs, response times, etc., are manifold.

Some mills, such as Milliken, have promised to share data developed in a project in this area. With the withdrawal of Sweden from the EC "BRITE" automated apparel project, there does not appear to be any serious project of model-building going on in Europe, nor is there evidence of collection of the data necessary to build such a model. If researchers at North Carolina State University can obtain support for their project at the School of Textiles, they may be able to produce the only working model of the total apparel system (from fiber to consumer). This would give them the opportunity to play "what if" games with the model to determine the effect of cost changes at various points in the system, the effect of consumer taste changes on all levels in the system, and many other variables which would give the industry an excellent planning tool.

The economic stakes are just too high for the apparel industry and its suppliers to give up without a fight. In Japan and England, even the retailers have realized the long-run benefits of maintaining viable domestic apparel manufacturing and are supporting their national programs. On the continent of Europe and in the U.S.A., the retailers are so busy succumbing to the greed of short-run excess profits they will not yet consider the long-run implications. The original Japanese timetable was optimistic and has had to be extended. Europe, because of British "hold-out" tactics and a general commitment to subsidize agriculture before considering anything else, delayed the funding of its program for two years. In the U.S.A., the Federal Government is aware of the situation but unable or unwilling to play a leadership or coordinating role. Large private firms are viewed with great suspicison by smaller companies and trade associations have not demonstrated the ability to lead or coordinate and academicians are busy sitting down together while they ask someone to stand up and supply some leadership.

Certainly some new technologies will be developed out of all this effort. The future of the apparel industry in the free industrialized nations will depend, in part, on what form these new technologies take. We may also expect to see a new organizational structure with new financial parameters evolve, for very few (if any) current apparel producers have the resources to undertake huge capital investment projects. Perhaps the traditional separation of the fiber/textile/apparel/retail industry segments will have to disappear before the total system can survive in the industrialized nations. Whatever the shape of the industry of the future turns out to be, it is apparent that massive changes in technologies, management systems, organizational structure, financial bases, and staffing requirements will be evolving during the next decade. The apparel business is not, therefore, a good place for cautious low-risk takers to be in the near future, but it does present an opportunity for high risk taking entrepreneurs who enjoy challenging environments.

American Society in the Age of Advanced Automation: Trends and Prospects

Julia C. Bonham
North Carolina State University

While technological change may seem to be the obvious solution to our problems of manufacturing competitiveness, adopting change is not accomplished without considerable impact on society. In this chapter, Dr. Bonham discusses the views both of those who advocate installing new technology as soon as possible, and of those who suggest that the possible societal impact of large-scale technological change make it imperative to plan not only for the change, but for the social consequences of that change.

Dr. Julia Bonham is Assistant Professor of History of Science and Technology, Division of University Studies, North Carolina State University. She received her Ph.D. in American Civilization from Brown University, and recently has completed one book and is working on another on the subject of historical technological change in the textile industry.

American Society in the Age of Advanced Automation: Trends and Prospects

Introduction

The high-technology revolution is underway. And the media rarely lets us forget it. In countless news stories we are told that the post-industrial society is here, and that the new technologies of the modern age are making their way into our lives at every point -- home, business, leisure, government, education, and even religion. Advanced automation is a graphic case in point. From the legions of personal computers residing in America's dens to the onset of computer-aided manufacturing, the influence of the new technology is spreading daily.[1]

This technological revolution is also a social revolution -- one that will affect all aspects of life, both individually and collectively. With so much at stake, it is not at all surprising that experts have been busily evaluating and forecasting the impact of advanced automation on society. However, although they have been speaking out at every opportunity in newspaper and magazine interviews as well as in scholarly treatises, they have been unable to agree among themselves on the quality of human life in a world increasingly populated with ever more sophisticated robots and computers. Instead, proponents of advanced automation and their critics are engaged in a heated media debate over the social consequences of the new technology.

Because the experts are polarized on this important social issue, an objective analysis of the human benefits and costs of advanced automation must take into account not only alternate perspectives but the impact of the debate itself on the process of technological dissemination. In this paper, the author presents the general arguments of both the automation proponents and the automation critics in an effort to establish the major points of contention and agreement between them. The author closes with a plea for cooperation, using historical precedents to highlight the possible negative implications of an intense, protracted public debate for the realization of both sides' goals.

The Pro-Automation Argument

Perhaps because its most vocal supporters are either vendors of the new technology or large industrial users, the pro-automation argument focuses almost exclusively on the application of the new technologies of flexible automation -- CAD/CAM and robotics -- in American industry. Proponents of advanced automation are quick to acknowledge that America's mature industries no longer compete successfully in world and domestic markets. They also are equally quick to point out the reasons for this sad state of affairs.

> Our friends from overseas. . . seem to make lazy circles up in the sky over the U.S. looking for our old, our stagnating, our complacent industries... then they come down for lunch... And it gets even worse if the competition happens to be a Japanese or other Far Eastern country whose pay scales are one-tenth to one-half those of an American company (sic).[2]

To address the related problems of foreign competition and expensive U.S. labor, industry proposes a technological fix: "We must try automating in this country, even if we have to beg, borrow, or lease to do it,"[2] for the only alternatives to automation are "evaporation" or "emigration".[3]

According to automation proponents, the new technology offers many advantages. Advanced automation increases productivity in manufacturing by decreasing man-hours and reducing the amount of materials, operating costs, and energy required to produce products. In the area of labor, for example, the industrial robot, priced at $40,000 or more, costs manufacturers between $3.80 and $4.80 an hour to own, operate, and maintain. With the human equivalent costing between $14 and $15 an hour (including fringe benefits), the cost advantage of the new technology seems obvious.[4] And when the relative productivity of human and machine labor is compared, the economic advantage of advanced automation widens. Capable of working around the clock, seven days a week, without lunch breaks or fatigue, the robot proves three to four times as efficient as the human worker, yet it never sacrifices quality due to boredom.[4]

Extending the discussion beyond the individual unit to flexible manufacturing systems, advocates of advanced automation report equally remarkable results. Priced at as much as

$25 million, it is difficult to believe that the installation of a flexible-manufacturing system would be cost effective, but industry representatives claim that it is. Relying upon statistical data obtained from Japanese manufacturers currently operating fully automated factories,[5] proponents claim that direct comparisons of starting-up costs between older plants and fully automated plants offer an inadequate guide to the incredible economies of flexible automation.

> Seventy-five percent of all machined parts today are produced in batches of 50 or fewer. Many assembled products too...are made in batches...
>
> In the past, batch manufacturing required machines dedicated to a single task. These machines had to be either re-built or replaced at the time of product change. Flexible manufacturing brings a degree of diversity to manufacturing never before available.
>
> The strategic implications for the manufacturer are truly staggering. Under hard automation the greatest economies are realized only at the most massive scales. But flexible automation makes similar economies available at a wide range of scales. A flexible automation system can turn out a small batch or even a single copy of a product as efficiently as a production line designed to turn out a million identical items. Enthusiasts of flexible automation refer to this capability as "economy of scope".[5]

When viewed in this context, advanced automation means cost-effective, flexible manufacturing. And since experts from all fields are predicting phenomenal growth in small markets which require short production runs, advanced automation seems to offer American industry the best chance to compete successfully against its Japanese counterparts.

> Burlington Industries, the nation's largest textile company, is learning to think small in its fight for survival against cheaper imports.
>
> Burlington has embarked on a $2 billion, 10-year modernization program (using robot fabric handlers and computer-operated vehicles), and Chairman and Chief Executive Officer William Klopman says the main goal is making the company nimble enough to compete in small markets.
>
> "Our customers want short runs," Klopman said. "They want flexibility. We're in the fashion business. Right now, they want variety and it's up to us to get set up to provide it in an economical way."[6]

The economic justification for advanced automation does not end with a discussion of increased productivity and industrial renewal. For, according to the pro-automation view, the benefits of the new technology will be felt in all sectors of the economy. Contending that the "...United States is enjoined in a struggle with Japan for its economic life and standard of living...,"[7] and that industrial competitiveness is essential to a favorable outcome in this battle, advocates offer advanced automation as a technological solution to the nation's economic plight.[8] Indeed, according to one proponent of industrial robotics, without "electromechanical slaves," another economic downturn and its accompanying inflation may well lead to the fall of American democracy.[9]

It is important to note here that state governments, and to a lesser extent the federal government, are committing themselves with increasing regularity to the pro-automation economic justification. For states in which ailing mature industries dominate the economy, the shift to high technology makes good sense. Governor Milliken of Michigan, for example, hopes to make his state the robot capital of the world, thus solving the twin problems of economic depression and high unemployment at once.[10] And his view is shared by others.

> Regional promoters from Florida to Oregon are spending tens of millions of dollars and enlisting figures from business, government and academia to attract investment in homegrown technology ventures. The idea is appealing: High technology is supposed to create jobs without causing pollution, is the wave of the future and brings in educated, high-income employees. Growth without pain. New versions of Silicon Valley and its most successful counterparts: The Research Triangle of North Carolina and Massachusetts' Route 128.
>
> Regional venture capital fairs...are held almost monthly from Atlanta to Monterey. State governments are spending anywhere from $2 million (Tennessee) to $30 million (Virginia) to start technology-transfer foundations to turn university research into marketable products... .[11]

If advocates of advanced automation are convinced that the new technology can help American industry halt its competitive decline and improve the economy, they are equally sure that it will improve American society by, among other things, ending high unemployment. In this view, layoffs in the United States are attributed to the importation of cheap products from abroad. "While unemployment spread here, Japan's 1982 trade surplus with the United States totalled some $12 billion, equivalent to 480,000 jobs lost by Americans to the Japanese."[12] Advanced automation, on the other hand, "protects jobs by increasing productivity and creates a demand for labor in the form of inspection, maintenance, and repair work."[13]

Automation advocates are confident that history bears out their conclusion that "[e]very new advanced technology developed has created employment."[12] Thus, they contend that it is irrational to believe otherwise.

> There is, of course, not a scrap of evidence in either theory or history to suggest that technological development won't increase employment and real incomes today just as it always has...
>
> One can go back as far as Adam Smith for evidence. In the first chapter of "The Wealth of Nations", Smith tells his famous story about the pin factory where a single worker without machinery "could scarce, perhaps with the utmost industry, make one pin a day, and certainly could not make twenty." But with the introduction of machinery a single worker could make as many as 4,800 pins a day. Given the number of pinmakers in England at that time and the demand for pins, such machinery should have created about 99% unemployment in the pinmaking industry. Obviously it did not because the vast increase in pinmaking productivity resulting from mechanization so lowered the cost of making pins that new uses for them were discovered. Moreover, the increased income of those who previously used high-cost, hand-made pins increased demand for other products and gave rise to employment in other industries, not the least being makers of machines for pinmaking.[14]

And if the weight of human history is not convincing enough, advocates of advanced automation clinch their position on employment by quoting labor leaders on the positive impact of technological advancements on productivity levels and our standard of living.[15]

The social benefits of the new technology do not end with an expanded job pool. In the pro-automation view, automated manufacturing also creates better jobs. Robots eliminate tasks that people "can't do, or won't do, or shouldn't do" because they are either dangerous or boring.[16] Indeed, Stanley Polcyn, vice-president of Unimation, Inc., believes that the elimination of assembly-line jobs should represent no hardship, for "it is a rare factory worker who really enjoys tightening the same bolt in the same car door for years on end."[12]

Of course, proponents recognize that the introduction of the new technology throughout the manufacturing sector of the economy will transform the present national job profile. For in the post-industrial society, there will be a proliferation of new jobs in the human services, technical services, and in "service industries yet to be invented."[12] Advocates take pains to point out that this transition from an industrial to a service society has been underway for some time, and that historically workers have voluntarily moved into service fields which, in their view, promise a higher standard of living and a higher quality of life. For example, Eugene Merchant, former director of research planning at Cincinnati Milacron, Inc., relies heavily on the work of the prominent sociologist, Daniel Bell, to support the claim that Americans prefer service work. He does so in order to establish a solid social justification for the dissemination of advanced automation technologies in American industry.

> The manufacturing industry, if it is to continue to exist and supply the manufactured products needed by the people of the world, has no choice but to move steadily toward realization of the automatic factory. That is in keeping with the common social good repre-

> sented by the above social trend (voluntary movement into service industries).[17]

According to this statement, then, to the extent that automated manufacturing will end high unemployment and make available to all Americans their preferred work in service jobs, industry's policy on automation is a socially responsible one.

Although a climate of optimism permeates the pro-automation vision of the future, advocates identify several potential deterrents to the realization of their economic and social goals through the application of new technology. Among these deterrents are the inability of the present educational system to meet the projected skills shortage, inadequate funding for research, middle-management and blue-collar resistance to technological change, and top-management uncertainty or reluctance to invest in capital-intensive improvements.

Of the many challenges facing our educational system, proponents of advanced automation seem most concerned about the predicted shortage of high-technology engineers and skilled technicians.[18] They contend that the existing system of higher learning will have difficulty meeting the demand for engineers and technical personnel trained to design, deploy, and service the flexible manufacturing technologies of the future. And the consequences of this predicament could prove dire.

> "When it comes to robotics, industry has available the equipment it needs -- now it needs the people to apply it." That assessment by the director of training for a large manufacturing company deeply involved with the application of robots puts the challenge in plain and simple terms. The shortage of engineers and skilled technical personnel capable of

> implementing and maintaining robots and associated equipment is a major stumbling block to sustained growth and utilization of this technology.[19]

Solutions to the potential shortage of high-technology personnel vary among automation proponents. Some industrial firms, especially the vendors of automation technologies, have pioneered their own training programs.

> General Electric is considered to have one of this country's largest and most advanced in-house robotics training programs and laboratory facilities. It's located at the company's Manufacturing Engineering Consulting and Applications Center in Schenectady, N.Y.
>
> An overview of G.E.'s approach to training comes from Donald L. McMann, manager of Training and Communications, G.E. Robot Applications Consulting. "We have a highly mechanized approach that's well documented with good material. Equipment-wise we have 20 robots in our lab and another six in a mobile education van which travels across the country and Canada. The lab also includes a complete line of programmable controllers which are taught primarily as they relate to robot systems."...
>
> McMann believes another strong point in the G.E. training effort is that the instructors serve in a dual capacity. They have expertise to serve as instructors and also as applications specialists. They are used interchangeably in these two roles as required.[19]

These industry-designed training programs, like the one at General Electric, have the advantage of allowing firms to train and update the skills of valuable, established employees with minimal disruption of the work schedule.[20]

Although automation advocates generally recognize that in-house training of skilled personnel is often desirable, they also look to places of higher education, vocational schools, and private educational organizations for help in the elimination of the skills-shortage problem. Progress in this area has been slow, but proponents report that these institutions and agencies are finally responding to the needs of industry.[19] For example, the National Center for Research in Vocational Education in Columbus, Ohio has conducted a project funded by the U.S. Office of Education to develop specifications and guidelines for two-year college programs in robotics and CAD/CAM technology.[19] In addition, the forty-member Education and Training Committee of Robotics International of the Society of Manufacturing Engineers "develops and coordinates educational materials and programs" and sponsors a certification program that "offers engineers and technicians formal recognition and documentation of their qualifications."[21]

Finally, Michigan has more than six community colleges, including the pioneering Macomb County Community College, actively engaged in designing and implementing robotics training programs.[19] Funding for these programs typically comes from private institutions and state - and federal - government agencies.[19]

The university role in the advancement of robotics and CAD/CAM is far more complicated than that of the community college or vocational school. For American universities are expected not only to build and equip training facilities for future generations of high-technology engineers, but to conduct theoretical and applied research in high-technology areas using expensive state-of-the-art laboratories.[22] There is general agreement that the level of funding will determine the universities' ability to fulfill both their educational and research missions. Thus, the competition for federal, state, and private funds is intense,[22] especially since the U.S. government has allocated some federal funds for its own research projects.[23]

To avoid the inevitable pitfalls of fierce competition among American universities for public funds, some proponents of advanced automation have called for a new national education policy -- namely, a high-technology Morrill Act, designed to "fuse a new partnership among federal, state, and local interests."[24] The keystone of this proposed legislative act, which theoretically modifies the 1862 land-grant concept to include high-technology engineering, is the matching grant challenge initiated at the federal level. Working on a 5-3-2 formula (five federal dollars, for three state dollars, for two industry dollars), this plan would raise funding for state-of-the-art facilities in the nation's three-hundred or so university-based engineering departments.

> This would not only add to the leverage of federal educational appropriations, but more importantly, it would ensure that federal allocations respond to the lead of industries and states in selecting those technical fields where the educational investment could be maximized.[24]

Some automation advocates believe that industry's research and personnel demands can best be met by direct university-industry collaboration.[25] North Carolina State University, for example, has reason to believe that this approach is beneficial to both its high-technology research programs and its high-technology training programs.

The Turbine Components Division of Westinghouse

> Corporation has provided a $100,000 robotic assembly to the N.C. State University Department of Electrical and Computer Engineering. (Dr. Wesley E. Snyder)... said that the assembly provided by Westinghouse will give faculty and students an opportunity to use an industrial robot in a laboratory environment for continued studies in the area of image processing systems.[26]

It is worthwhile noting here that university-industry cooperation has the blessing of the federal government.

> The U.S. General Accounting Office (GAO) has given the university (North Carolina State University) high praise for its efforts in promoting university-industry interaction.
>
> In a report prepared for the joint U.S. Senate-House Committee on Science and Technology, the GAO reported that NCSU is heavily involved in all three types of university-industry collaboration -- research parks, cooperative research centers and industrial extension services.[27]

If outdated educational facilities and insufficient training and research programs threaten the smooth transition to post-industrialism, then the reluctance of top management to invest readily in the new technology affords another danger. Realizing that the multi-million dollar price tag for a flexible manufacturing system tends to slow the technology's acceptance rate significantly, some proponents explain that this stumbling block can be partially overcome by establishing "islands of automation" -- "a robot here, a computer there, surrounded by older automated equipment or workers."[28] Companies which embrace this piecemeal approach must still determine the most suitable equipment for their plants, and top management rarely enjoys the expertise required to make such complicated decisions.[29] For firms considering more radical modernization, decision-making becomes even more difficult, for management must justify not only the initial cost of the new machinery but the major disruption of the manufacturing process. "Few plant managers are eager to rip up assembly lines to install equipment, particularly when consultants may want 16 weeks to design a new system."[28] Despite the expense and disruption, proponents warn that there is no viable alternative.

> ...Concentrating on cost reduction at the expense of investment for the future could make matters worse a few years from now, some businessmen say. "Just by cutting costs, we probably could get a pretty good run of profits for several years when business improves," says James T. Bartlett, president of Cleveland Twist Drill Co., a subsidiary of Acme-Cleveland Corp. "But if we don't also adjust technologically, when the next downturn comes we'll be in a valley so dark you couldn't believe it."[30]

To minimize the negative impact of top-management uncertainty on the dissemination process, both universities and vendors of the new technology are developing programs to help potential buyers choose the best path for them.[31] For instance,

> [a] new manual being prepared at the NCSU (North Carolina State University) engineering school will help teach companies how to compute the benefits of installing the expensive new equipment. Robots, for example, can cost $60,000 -- often only a quarter of what it costs to adapt them to other assembly-line equipment.[28]

Ardent proponents of advanced automation, General Electric foremost among them, hope that such programs, along with a focus on specific applications development, will reverse the 1982 trend which lead one executive to observe that potential customers "aren't lining up outside our door."[32]

Middle-management resistance to technological change offers still another challenge to industrial executives, specifically, and to the goals of the automation advocates, generally.

> Both literature and interviews reveal that employer representatives, particularly at middle management level, more often constitute the real barrier to the introduction and effective use of technological innovations.[33]

Advocates of automation acknowledge that this problem must be studied and solutions found before the promise of the new technology can be realized.

> We must start with Plant, Assistant Plant Manager, Operations, and Engineering Manager level. If this group is not fully informed and does not fully agree to take an active part in the planning, installation and launch of the first application, then this is the point to stop. [34]

Despite a dearth of formal studies devoted to the problems, some advocates have suggested possible reasons for this unexpected reaction from middle managers. Convervatism among "production people" is one popular explanation.[35] Another concerns a perceived loss of authority on the part of middle managers as the organizational status quo is threatened.

> Ivan Vernon anticipates the future manufacturing organization and the skills required to manage it. Among the points he stresses is the potential loss of the manufacturing manager's autonomy. "Indeed, he may become just another member of a team and perhaps not the most important one."
>
> No one yields power and authority willingly, and, faced with this probability, it is predictable that management will try to install robots without disturbing the existing hierarchy.
>
> However, authority will need to be vested with those with knowledge rather on hierarchical position. As operators are trained and become skilled workers, their level of knowledge is likely to exceed that of their supervisors. Supervisors can be expected to resent this threat to their authority and may be prone to watch the robot fail, welcoming a return to manual operation. The probability is increased by the fact that the supervisor's rewards are based on getting the product out, not on enhancing the sophistication of the operation.[35]

According to the pro-automation group, executives can help eliminate resistance for these reasons by demonstrating sensitivity to the plight of its middle managers.

> . . .[T]op management must demonstrate leadership which enables subunit managers to take a fresh look at the various stakes each has in the installation. Top management must create the climate in which their managers may jointly examine costs and benefits to their self-interest. Innovative solutions to integrate disparate goals will be time-consuming and may appear, at first glance, to be less productive than more tangible measures usually associated with manufacturing.[35]

Most automation proponents seem painfully aware that discontent among displaced or obsolete blue-collar workers represents a serious threat to the smooth and swift adoption of advanced automation in industry.[36] Typically, their response to their potential deterrent takes one of two forms. Some attempt to redirect attention away from technologically induced unemployment, thus minimizing the amount of bad publicity devoted to it. Consider this statement by Joseph Engelberger, founder of Unimation, Inc.

> Recently I argued with a social scientist on television, who said, "Hey, 128 robots just went to work at Chrysler. Haven't they just taken away 256 jobs?" I replied, "Yes, but there are 600,000 people out of work as a result of a depressed automotive industry. They did not lose their jobs because of robots; they lost their jobs because foreigners, notably the Japanese, have been able to make better cars and sell them here at lower prices." If we do not use the most advanced productive techniques, there will no longer be an automotive industry in the United States. We will be reduced to servicing Japanese automobiles.[37]

Other advocates begin with the contention that, historically, technologically induced unemployment has proven to be a temporary outgrowth of automation, and as such a necessary short-run evil endured by the few for the long-term good of the many. In this view, the goal of society should be to minimize the deleterious effects of the bumpy transition of post-industrialism for these unfortunate few.

> All this isn't to say that some specific individuals may not become unemployed as a result of technological change. Legitimate efforts should be made

> to ease their plight and help them find new work. But such instances shouldn't obscure the larger good to society from increased productivity.[14]

When proponents are asked to describe exactly what kinds of efforts should be made to alleviate suffering and by whom, differences of opinion emerge.

For some, nothing needs to be done because the solution to worker displacement can be found in the existing system.

> At Chrysler we have some 80 robots compared to 8,000 people. That's a very small percentage to have to be retrained. And this (introduction of robots) has occurred over several years. So the replacement of an employee can very easily be accommodated by not hiring someone for a job that comes open through attrition or any other way.[38]

However, a growing number of automation proponents are of the opinion that direct action is necessary to lessen the human costs of technological change, and some of these believe that industry must take an active role in programs to alleviate the suffering that comes with dislocation. Citing the Japanese as a model,[39] some recognize that job security and retraining programs must be a part of any successful effort to undermine blue-collar resistance. For example,

> GE already has guaranteed considerable job security for workers affected by automation, (executive vice-president James) . . . Baker says. . . "We are willing to retrain wherever we can, or to help ease the move from unskilled to skilled jobs."[32]

Other proponents believe that top-management sensitivity to blue-collar fears can go a long way in smoothing the road to advanced-automation manufacturing. For example, the Digital Equipment Corporation "ensured the success of its experimental Robot Research Project with "heavy doses of sensitivity, honesty, and humor."[40] Members of this more liberal sub-group cajole industry to embrace the three laws of robotics.

1. Organizations may not install robots to the economic, social, or physical detriment of workers or management.

2. Organizations may not install robots through devious or closed strategies which reflect distrust or disregard for the work force, for surely they will fulfill their own prophecy.

3. Organizations may only install robots on those tasks which, while currently performed by men, are tasks where man is like a robot, not the robot like a man.[33]

Finally, some advocates are finding joint industry-government-union ventures a positive approach to handling certain social problems, because working together lessens the traditional tension between management and organized labor and thereby facilitates mutual understanding and goodwill. For example, by including the ACTWU in a revolutionary R & D project entitled the Textile/Clothing Technology Corporation, industry participants in the program are optimistic about their future relations with one of the major textile unions.[41]

Although the pro-automation group may not be able to agree on the best way to re-introduce technologically displaced laborers into the work force, they do not appreciate being the brunt of criticism for this reason or any other. For example, some advocates are quick to explain exactly how critics fail to present an accurate view of the social impact of high technology. First, critics "tend to generalize from particular case studies." Second, they "tend to be overly optimistic about the speed at which new technology will be adopted." Finally, they overlook the historical fact that "the new technology is best regarded as a development from previous technologies."[41] As an evolutionary development, it is not likely to cause severe job loss or other radical social change. In other words, there will be time to adjust to new conditions.

> Now whenever you get into an argument about whether the people are going to fit the environment or the environment is going to fit the people, you have to ask yourself whether both sides of that equation may not be adjustable -- whether the actual spectrum of jobs a society provides does not, in fact, adapt itself to the level of human intelligence in that society.[42]

Since both the environment and the people are adaptable in this view, unemployment becomes a "non-problem."

Some proponents of advanced automation are content merely to chide their critics. Others take a far more aggressive stance, often likening their opponents to the machine-smashing Luddites of the first Industrial Revolution.

> The Luddite argument has no more validity today than it had in 1811. Those who dwell on the apparent adverse effects of technological change on employment are only creating unnecessary fear and anguish among workers who worry they will be laid off. Though they may not in fact advocate restrictions on the introduction of technology, as the Luddites did, they may be sowing the seeds of discontent which could take root in other ways. Already there is pressure for trade restrictions, industrial

> policy, plant-closing laws and government bailouts for companies in declining industries. The new Luddite argument only adds to the pressure.[14]

These proponents believe that criticism is counterproductive and even harmful to the already vulnerable blue-collar workers. Thus, in addition to being alarmist, criticism of the pro-automation argument is somehow un-American.

The pro-automation argument is an impressive one. Relying heavily on historical evidence, statistical analysis, patriotism, and capitalist ideology, it offers an optimistic vision of the future which promises economic growth, better jobs, and a higher quality of life. But however appealing the pro-automation view is in its conceptualization and presentation, it tells just one side of the story. For there is another group of experts who prefer to challenge the arguments and conclusions of their pro-automation counterparts, while engaging in some social forecasting of their own. Ranging from ardent condemnation to support with reservations, the critical responses, when taken together, conclude that advanced automation is not the panacea that its supporters believe it is.

The Critical Response To Advanced Automation

Perhaps more than any other element of the proponents' vision, the pro-automation depiction of the relationship between technological dissemination and employment patterns engenders the greatest critical response. Indeed, critics find fault with this feature of the pro-automation position on several counts. For example, even the most moderate critic, engineer Martin Goland (President of Southwest Research Institute), believes that the pro-automation contention that the new technology will yield more jobs than previously available is questionable.

> [Interviewer] I take it that the postindustrial futurists, as you refer to them, envisioned the day when one of the maxims of technological history would be refuted, namely, that advances in technical sophistication always create more new jobs than they render obsolete. This apparently has always been the case until now.
>
> [Goland] Yes, this is so. And I agree with them, The new element that has been added to the equation is our ability to produces [sic] semi-intelligent machines capable of making the same kinds of decisions that the production worker is called upon to make. The worker thus becomes redundant to a machine which works continuously for three shifts if necessary and performs its tasks completely accurately. This is a capability made possible by virtue of the computer and/or microprocessor, which, amazingly enough, entered upon the scene only within the past decade.
>
> [Interviewer] But more sophisticated machines take many skilled people to design, to produce, and to service and maintain them.
>
> [Goland] Correct! But the interesting and most meaningful question is whether or not the jobs added along these lines nearly equal the numbers of jobs that have been made redundant. . .[43]

Using statistical projections to support their claims, critics suggest that the number of new jobs generated by advanced automation probably will not compensate for the number lost due to automation. For example, Robert Ayres and Steven Miller of Carnegie-Mellon University contend that robots may eliminate two million jobs in the metalworking industry alone by 1990, while industrial expansion will provide just one-half million replacement positions. And the problem does not stop there, for "job losses that may have a minor effect on total employment can have a magnified impact if they are concentrated in a specific industry."[44]

Timing is another factor of concern to the critics. How long will it be before automation-induced layoffs swell the ranks of the unemployed in the United States? Skeptics claim that automation-induced layoffs are part of the present socio-economic situation and not something to worry about in the distant future. In their view, however, the media and thus the American people do not realize this because recent economic instability has obscured the deleterious impact of automation on unemployment figures and blurred the relationship between structural unemployment and automation-induced unemployment.[45] Fred Guterl, Associate Editor of IEEE SPECTRUM, uses the case of the John Deere & Company to explain the situation.

> The depressed state of farming has had a dramatic effect on Deere's production and employment. In 1979 the company made 186 tractors a day; now it makes 94, according to David Neal, a spokesman for the local chapter of the United Automobile, Aerospace, and Agricultural Implement Workers of America (UAW). This cutback has obscured the social effects of automation and diverted the attention of unions to economic problems. The UAW has 12,000 (sic) members at Deere's Waterloo complex in 1980, according to Mr. Neal; now it has about 6900. A total of 4500 Deere workers were laid off, and 600 others either quit or retired.

> Most job losses of UAW members were due to the production cutback. When production gets back up to full swing, many of these workers will be rehired, according to Troy McAfee, manager of manufacturing at Deere's Waterloo Component Works. But meanwhile, computer-driven automation has eliminated many jobs in the complex, even though automation has not been directly responsible for any layoffs. The automated storage and retrieval systems, in particular, eliminated the need for forklift operators and others involved in materials handling and transport at Waterloo. "The loss of jobs has been tremendous--more than anyone expected," said Mr. Neal. Before the new systems were built, 60 workers were needed for materials transport; today 14 load and unload parts, according to Mr. Neal.[45]

This confusion over automation-induced layoffs versus structural unemployment is clearing up fast. With increasing regularity, the media has been referring to the impact of automation on present-day unemployment figures, while describing the human suffering which results when automation transforms temporary unemployment into permanent joblessness.

> Evelyne Mills is one of the long-term jobless for whom the recovery isn't producing much hope. She is angry at the system and at Allied Corporation's Bendix machine-tool unit (sic), a Cleveland company that laid her off more than a year ago (1982) after 10 years of service. Her anger was inflamed early last month when Bendix held a reception and introduced its new computerized machining centers, imported from Japan, right in one of the plants where the company is still making cuts to reduce its 1,700 workers to 600. As officials demonstrated the

> machines for guests inside the plant, she and about 25 former workers demonstrated outside, protesting being left behind by imported technology and exported jobs. "I was a skilled worker," she says, "a plant inspector for seven years. It's hard to believe no one wants those skills now."[46]

Two lessons emerge from this story. The first concerns the relationship between capital expenditure, labor, and technological innovation in difficult economic times. Whether the problem is inflation or intense foreign competition, technology and labor under these conditions compete for whatever capital is available. According to the critics, when management believes that it must choose between labor and modernization programs, labor tends to suffer. Even industrial executives acknowledge this trend.

> The job losses are continuing. . .(Barberton, Ohio's) largest employer, Babcock & Wilcox Co. . . cut 230 employees this year (1983). The second-largest employer, PPG Industries Incs.'s industrial chemicals plant, says more than 700 employees will lose their jobs by year-end. This plant was built here in 1899," says Robert Congleton, a PPG executive. "We have modern plants in Louisville and West Virginia, and their operations are curtailed now. If there's an upturn, it's the modern facilities that will handle it. We had to make a decision about where to put our capital."[46]

Second, the story calls into question the credibility of the pro-automation argument that skilled workers of all kinds will be in ever greater demand. Indeed, some critics have argued that:

> . . ."we've peaked in terms of our need for the classical technician. The machinery is more and more self-diagnosing. There's less of a judgement factor. The computer spits out an instruction to change such-and-such component, and you do it. We can upgrade a $5-an-hour assembler to a $6- or $7-an-hour junior technician with little advanced training. A few years ago, we were recruiting skilled technicians as aggressively as we're recruiting engineers today. We're not doing it now. Organizations that train technicans should realize that they're preparing people for a smaller and smaller marketplace.[47]

In addition to disputing the pro-automation argument regarding the relative numbers of jobs to be lost and gained with the introduction of advanced automation into the factory, critics also question the pro-automation contention that robots will improve the quality of work in factories. Indeed, they argue that there is a real possibility that the quality of the work experience will deteriorate as the new technologies enter industrial settings in greater numbers.

> On the negative side, from labor's standpoint, are the fewer people on flexible lines. And because, as in automobile assembly lines, the flexible line must be kept moving, breaks are staggered and there is reduced interaction among workers on the line, who tend to be isolated, according to Mr. Neal. . .
>
> [Moreover], [a]lthough computer-drive automation gives the worker more responsibility and autonomy, it also gives him less control over the process he is tending. The loss of a sense of control can make jobs as machine tenders undesirable, according to Robert T. Lund, senior research associate at the Center for Policy Alternatives,

> Cambridge, Mass. And since computers give managers easy access to data, many workers are reportedly anxious about their performance being monitored, he said.[45]

To support such claims, critics offer evidence taken from existing case studies. For example, at the Robotics Institute at Carnegie-Mellon University, researchers conducted a prototype study of worker reactions to the introduction of a robot into a machine shop.

> When the robot was introduced, a manufacturing cell. . . was created. The robot provided material handling functions for two milling machines. An operator was then responsible for the two milling machines and the robot. The introduction of the robot removed the materials handling activity from the operator's jobs and added a new activity, robot operator.[48]

Workers found that with the switch from manual operations to cognitive activities (monitoring) came increased responsibility and isolation. "I don't have time to talk to anyone. . . .I don't want them breaking my concentration. . . .I'm isolated now."[48] Indications of worker stress followed. "It's nerve racking. . . there are lots of details. . . It's an expensive piece of equipment."[48] With respect to the stress, the researchers commented that

> [t]here was much speculation during our first visit to the plant about whether an operator who was particularly quick would be able to beat the robot. By our second visit, workers seemed resigned to the fact that the robot would always be able to outproduce a human worker. . .

> Objectively, after the robot introduction, the operators controlled the robot and the two milling machines. However, operators subjectively viewed the situation as one of competition between them and the machine. This may be another source of stress.[48]

More interesting work, theoretically, might balance the negative impact of increased stress. But researchers reported that at least one operator did not find his new job more satisfying than his old one.

> One of the robot operators in our sample mentioned that although he found observing and monitoring more boring than manual activities, he was currently satisfied with his job because there were still many manual set-up activities. The operator commented, however, that without these set-up activities, the current job would be more boring than his previous job. We think this incompatibility between activities required by the job and preferences of the worker was another source of stress.[48]

According to Harley Shaiken, labor and technology analyst at the Massachusetts Institute of Technology, the finding of the Robotics Institute study should be taken seriously. In his report to the U.S. Office of Technology Assessment on the IMPACT OF PROGRAMMABLE AUTOMATION ON THE WORK ENVIRONMENT, Shaiken states that the introduction of computerized machine tools into several small metalworking shops leads to greater "centralized control of shop operations", decreased "overall skill requirements", and increased complaints that jobs "had become routine and boring."[49]

As a final critique of the pro-automation depiction of the

relationship between technology and human labor, the critics offer a warning: flexible manufacturing systems may limit opportunities for advancement.

> . . .[T]he opportunity for advancement--an important factor in motivating workers--is bleak in factories that use flexible manufacturing, according to Mr. Lund, at the Center for Policy Alternatives. Since they require less labor, they require less supervision. And those supervisors needed are usually technical types with training as engineers or technicians.[45]

Although critics tend to focus on the potential negative impact of advanced automation on the workplace and labor force, they are quick to point out that problems such as stress and increased isolation can be costly for the organization as well. For example, in concluding their study of "the human side of robotics," researchers at the Robotics Institute suggested that

> [i]t would be premature at this point to speculate whether this increased stress is good or bad for the individual or the organization. . . .[but] [s]tudies have shown that increased stress is associated with increased turnover and absenteeism and that stress can lead to either increments or decrements in performance.[48]

If the effect of worker stress on shop efficiency is in question, so is the impact of worker isolation on productivity. In the following case, the "workers" are a product designer and a machinist, who have been asked to produce a product designed by computers.

> The temptation is to accept the computer model

in place of the reality of the shop floor. The two, however, are not the same. When computer-aided design is used to enhance the close relationship between the designer and the machinist, it expands the possibilities of production, but the computer-aided design as a replacement for that interaction poses some real dangers. It becomes very easy to lose the feel of making things, the intangible experience that comes only from close contact with the realities of production as they take place.

An example of this loss occurred in a British aircraft factory. A young designer, using the latest in computerized equipment, designed an igniter for a gas turbine engine. The igniter was then made on a computerized machine tool. It was a brilliant design, but somewhere in the process the decimal point was moved accidentally one place to the right. The computerized instructions came down to the shop floor, and the worker involved built the igniter according to the specifications. The only problem was its size: 10 times the projected design. When the worker carried it up to the young designer in the shop office, the designer looked at it and did not immediately notice any problem with the part. . . How do manufacturers compensate for these mistakes when the designer has become severed from the realities of the shop floor?[49]

If the critics have grave reservations about the quality of the work experience for those who will be employed in automated factories, they are even less sanguine about the prospects of unskilled laborers in the high technology environment. Blue-collar workers are, after all, made obsolete by the new technology.

> "Four million Americans out of a job by 1990," blares one forecast. And according to Harley Shaiken. . .in the next ten years General Motors alone will purchase 20,000 robots. These machines could displace another 40,000 to 50,000 auto workers besides the thousands already laid off directly and indirectly by automation.
>
> "The argument is often raised," Shaiken says, "that unless we introduce robots, industries in the United States will not be able to compete, and even more jobs will be lost. But that's half an argument. It doesn't tell us who will lose their jobs. For workers, robots or foreign competition is merely a choice between hanging and the electric chair. And we, as a society, will have to break out of that bind."[50]

Without action to redress this social trend, critics conclude, a whole segment of our society will be condemned to a life of despair--something akin to the existence of the structurally unemployed today. And their world is one of depression, family breakdown, violence, and alcoholism.[51] The fact that minorities and middle-aged blue-collar workers are more susceptible to layoffs than ever before only serves to make the social consequences of prolonged joblessness all the more unacceptable in the eyes of the critics.[52] In the face of this physical, mental, and moral erosion of American labor, American society must find answers to several difficult questions: Will the natural process of attrition provide adequate numbers of jobs for workers who are displaced by automation? Is retraining the answer? Or relocation? Or job security agreements between labor and management? Do we have adequate mental health services to help the displaced? Who should pay for these social services? Is industry committed to providing social services to help the unemployed?

Certainly, critics are not convinced that attrition alone will solve the problem of automation-induced unemployment.

> It is often noted that technological displacement would be minimized if the rate of robot introduction were paced by the attrition rate. . . An examination of industry attrition rates and age distribution of manufacturing operatives and laborers suggest this strategy is not feasible. According to the Bureau of Labor Statistics data, only one to three percent of the workforce in metalworking (SIC 33-38) leave their place of work as a result of quits, discharges, permanent disability, death, retirement, and transfers to other companies.
>
> [Moreover,] [c]ontrary to the notion that many manufacturing workers are old and nearing retirement, the vast majority of the manufacturing workforce still has 20 or more years of active worklife ahead of them. As of 1980, between two-thirds and three-fourths of operatives and laborers were less than [sic] 45 years old, which means that barely a third of the workforce would be retired in the normal way by the year 2000. . . . On the average, skilled workers are older, but they are not as likely to be replaced by robots in the near future.[53]

Retraining and, if necessary, relocation are possible solutions, especially since almost all critics and many proponents of advanced automation seem to support this approach. Certainly the unions, after a slow start,[54] are indicating that retraining and other issues involving automation will increasingly become the main points of collective bargaining. Discussions will focus on establishing procedures for technology assimilation that will minimize human misery during the transition. Union leaders have suggested the enactment of a "technology bill of rights" to force management to notify unions prior to the introduction of new technology and to encourage management to share the profits accruing from increased productivity with their employees through shorter hours, higher wages, paid personal holidays, transitional allowances for the permanently laid-off, severance pay, and corporate-sponsored retraining programs.[54] Unions have also been making headway in the area of job security.[55]

Although critics typically support union efforts to establish programs which address the problems of the long-term unemployed, they seem equally sure that union action alone is not enough. Ayres and Miller, for example, conclude that the prevention of "social trauma due to rapid introduction of robotics, without impeding technological progress itself" demands:

*Identification of vulnerable categories of workers well in advance of actual job elimination.

*Long range planning by industry and government for the future employment needs and new job skills requirements.

*The provision of effective education and training facilities to upgrade workers from skill categories that are, or will be in surplus supply to skill categories that are scarce.

*The provision of effective facilities to locate suitable jobs and place workers in them, with relocation assistance if necessary.[53]

In this view, then, the level of suffering endured by

working Americans during the transition to post-industrialism depends upon the level of commitment by industry, government, and the universities. Some critics of the pro-automation vision of the future are not certain that these institutions are committed to the kinds of policies and actions necessary to ease the plight of American labor and improve society as a whole. In the case of higher education, for example, there is certainly a push on to meet the demand for high-technology engineers. But are engineering students today being prepared at the university to be socially responsible citizens? According to Frank Rose, the answer is no. "The future is being designed by engineers who have been trained simply to be engineers. The one thing we know about the future is that it's more complicated than that."[56] Indeed,

> . . .a postindustrial society requires more than just a lot of engineers. It requires engineers with the vision and the forcsight to scc where they are going and the verbal capacity to tell us about it. The real problem with engineering students is not that they are driven by the desire to make twenty-five thousand dollars a year--that's just their way of putting the quest for self-fulfillment in quantifiable terms. The problem with engineering students is the problem with everybody else in America: they can't seem to bridge the two sides of knowledge. Our institutions, in other words, are not developing the integrative functions required for the thoughtful application of complex systems.[56]

If the education of engineers concerns some critics, then what appears to be industry's shaky commitment to its workers worries many more. In their "search for excellence" in corporate America, Thomas J. Peters and Robert H. Waterman, Jr. found that humane corporate personnel policies not only benefit individual workers but enhance productivity. Nevertheless, many companies practice a cruel and costly form of hypocrisy.

> Treat people as adults. Treat them as partners; treat them with dignity; treat them with respect. Treat THEM -- not capital spending and automation -- as the primary source of productivity gains. These are the fundamental lessons from the excellent companies research. In other words, if you want productivity and the financial reward that goes with it, you must treat your workers as your most important asset. . .
>
> Genuine people orientation is in marked contrast to two major alternatives all too often seen in companies: the lip service disaster and the gimmicks disaster.
>
> The lip service disaster is arguably the worse of the two. Almost every management we've been around says that people are important -- vital, in fact. But having said that, they then don't pay much attention to their people. In fact, they probably don't even realize their omissions. "People issues take up all my time," is the typical rejoinder. What they often really mean is, "This business would be so easy if it weren't for people."
>
> Only when we look at the excellent companies do we see the contrast. The orientation toward people in these companies often started decades ago -- full employment in times of recession, extraordinary amounts of training when no training was the norm, everybody on a first-name basis in times much more formal than ours, and so on. Caring runs in the veins of the managers of these institutions. People are why those managers are there, and they know it and live it.[57]

Peters and Waterman suggest that executives often do not realize that they have adopted a less than progressive position. Whether they do or do not, Ayres and Miller suggest that the "uncommunicative attitude" of top management about proposed technological innovations in the plant means that "very little is being done to warn or prepare those workers whose jobs may be eliminated, or substantially modified as a direct result of introducing robots."[53]

The case against industry does not end there. Consider the evidence below. In recent years, observers have recorded a spate of corporate-sponsored retraining program failures.[58] Moreover, the president of Burlington Industries has gone on record saying that his textile workers can not be retrained and relocated.[59] Finally, there is the increasingly popular practice in industry of telescabbing. Both manufacturers and unions are discovering that automation can be an effective strike breaker.

> A strike against Consolidated Edison Co., which ended last week [August 1983] after more than two months, went unnoticed by most New Yorkers because the electric system is highly automated. A major strike against several oil refineries had little impact because the pumps and valves are almost completely automated. And a strike at a jet-engine plant in Ohio didn't stop production because computers have simplified the work of highly skilled machinists so it now can be done by superiors and secretaries. A new study by the Wharton School at the University of Pennsylvania found that many capital-intensive companies achieve 90 percent of usual production levels during a strike; most other firms manage 50 to 80 percent.[60]

In the view of some critics, the decline of unions comes at a most inopportune time, for unions have been the traditional defenders of the wage earner in the United States.[61]

If, as some critics suggest, industry's commitment to its employees is questionable, then what about the federal government's role in finding solutions to the human problems created by advanced automation? In media reports over the past few years, the federal government has not looked good. In a WALL STREET JOURNAL article on "how to help 'displaced workers,'" for example, readers were told that their government has been refusing to establish an effective national job-matching system for years.[62] Moreover, after Congress had allocated more than $100 million to help retrain and relocate displaced workers in 1983, a WALL STREET JOURNAL reporter explained to his readers that the U.S. Department of Labor had other plans for the money, so that a federal judge was forced to order the administration to comply with the wishes of Congress.[63]

Although their criticisms of industry and government focus on the plight of the blue-collar worker in the age of automation, critics find even more reason for concern when the discussion is extended to include white-collar workers as well. When all service positions from executive to menial labor are considered, they argue, the social implications of advanced automation become all the more profound. After all, office work is already succumbing to computerization.[64] It is the same for sales. Not even management is immune.

> The jobs that are not being automated involve simple structured decisions, . . . but many managers and engineers make more complex decisions that may some day be automated by more sophisticated computers. The fifth-generation computer program, in which artificial-

> intelligence techniques are being developed, may mark the beginning of automation of white-collar jobs. . .[65]

Replaced by machines, both blue- and white-collar workers, in industry and outside of industry, must seek employment elsewhere. According to proponents of advanced automation, most of these people will find desirable jobs in new service industries. Critics agree that most replacement positions will be service areas, but the jobs available in the greatest numbers will be far from appealing.

> The widely held expectation is that a dazzling array of new occupations and skills will accompany the increase of computers in the industrial work place. . .
>
> This rosy view is challenged, however, by the darker vision: the creation of a pyramid of skills that concentrates a few creative and meaningful occupations at the top, while the rest wind up with fewer skills and subject to new forms of monitoring and electronic control. David Rockefeller, Chairman of the Chase Manhattan Bank, and hardly the premier U.S. social critic, has underscored these apprehensions by predicting that present forms of automation may lead to a "two-tier society, with satisfying and well-rewarded work for some, while the rest are left to grapple for unskilled jobs." Not only will skills be eroded, according to this view, but managerial control will be increased and the work environment will become more and more authoritarian.[49]

In this view, the United States is destined to become a two-tier or two-class society in which the smallest class is very rich and the largest class is comprised of low-paid menial workers who do not want the jobs they are forced to take.[66]

What is conspicuously missing in this scenario, according to Bob Kuttner, is the backbone of American society--the middle class.

> There is a good deal of evidence that job opportunities in the United States are polarizing, and that, as a result, the country's future as a middle-class society is in jeopardy. What the decline of the middle class would mean to the country can only be guessed at, but it presumably would be unwelcome to the millions of parents who hope that their children can move up the economic ladder; to American business, which needs a middle class to consume its products; and to everyone who is concerned about fairness and social harmony.[61]

Included in the body of evidence that Kuttner uses to support his contention that the middle class is shrinking are figures taken from the U.S. Department of Labor Statistics.

> Statisticians at the Department of Labor have been comparing the salaries paid in the twenty fastest-growing occupations with those in the twenty that are in the sharpest decline. . . Ronald Kutscher, an assistant commissioner of the Bureau of Labor Statistics, says that workers in the fastest-growing industries are earning, on average, $5,000 a year LESS than workers in industries that are declining or that are growing most slowly.[61]

Since the industries with more jobs are on average paying less to their employees than large industrial employers had in the past, Kuttner concludes that the redistribution of wealth that will usher in the two-tier society is presently underway.

If this threat to the middle-class standard of living is real, then the pro-automation vision of the future, which promises to preserve the American way of life, can not be realized. Consequently, some critics have accused automation enthusiasts of deception and have challenged the pro-automation argument on moral grounds. They have even suggested resistance as an appropriate political response to the situation. Consider this comment from writer Thomas Pynchon:

> "If our world survives, the next great challenge to watch out for will come - you heard it here first - when the curves of research and development in artificial intelligence, molecular biology and robotics all converge. Oboy. It will be amazing and unpredictable, and even the biggest brass, let us devoutly hope, are going to be caught flat-footed. It is certainly something for all good Luddites to look forward to if, God willing, we should live so long. Meantime, as Americans, we can take comfort, however minimal and cold, from Lord Byron's mischievously improvised song, in which he, like other observers of the time, saw clear identification between the first Luddites and our own revolutionary origins. It begins:
>
> AS THE LIBERTY LADS O'ER THE SEA
> BOUGHT THEIR FREEDOM, AND CHEAPLY,
> WITH BLOOD, SO WE, BOYS, WE
> WILL DIE FIGHTING, OR LIVE FREE,
> AND DOWN WITH ALL KINGS BUT KING LUDD![67]

Marxist social commentators present an even darker picture. But unlike Pynchon who faults government leaders, these critics believe that the application of new technologies must yield social injustice because industrial capitalists rarely exhibit genuine concern for their human laborers. Historian David Noble explains:

> The drive for total automation is promoted in the name of patriotism, competitiveness, productivity, and progress. Its twin aims, however, remain control and domination, and the extravagance of the effort is only matched by its absurdity.[68]

According to Noble, to eliminate the social injustices of industrial capitalism, organized resistance and more is necessary.

> The problem is political, moral, and cultural, as is the solution: a successful challenge to a system of domination which masquerades as progress. Such a challenge will require opposition to technology in its present form -- to buy time and cripple the current attack [by technophiles like a pro-automation group]. And it will require political mobilization and vision, cultural inventiveness and rejuvenation, and a revitalization of moral confidence. But it will also require once and for all a transcendence of the irrational and infantile ideology of technological progress which has confounded Western thinking for at least two centuries -- an ideology which has for too long obscured the realities of power in society, provided legitimation and cultural sanction for those who wield it, and paralyzed any and all opposition against them.[68]

Even though the Marxist rhetoric may seem strident and overstated, many critics believe that without careful study and planning to mitigate against the worst features of post-industrialism, social dissension and even violence may well result.[53]

****

If accurate, the critical response to the pro-automation argument is disturbing, for it portends a society far removed from the American ideal based on a strong middle class and on the potential for upward mobility. Whether or not the critics are right, however, the presence of such determined opposition to the pro-automation vision of the future is significant because it means that the people who should be able to forecast the human costs and benefits of advanced automation -- the experts -- are unable to reach a consensus opinion. Thus, all the rest of us must conclude that the United States is undergoing this process of social transformation without a clear end in sight. Even if this observation is accurate, there still may be reason for cautious optimism.

Trends and Prospects

In comparing the pro-automation argument to the critical response, one must conclude that the major point of contention between them concerns the relative importance that each imparts to two issues: the revitalization of American industry and the social implications of advanced automation. Automation advocates emphasize the do-or-die position of American industry in this period of intense foreign competition. In doing so, they acknowledge that advanced automation will cause some localized hardships, but explain that the human dilemma will be far worse if industry is forced to practice labor-intensive manufacturing processes to the point of extinction. Critics, on the other hand, fear that the human costs of automation will be overlooked in the rush to rescue our struggling industries. Thus, they demand that the private sector and the Federal government accept their responsibility to displaced workers and the American citizenry at large by taking positive action to minimize social turmoil.

A cursory look at these positions indicates a clear division of thought. But a closer examination reveals the presence of several common assumptions and conclusions. After all, whether explicitly or implicitly, both viewpoints acknowledge that the present state of American industry is not desirable and that change is in order. Furthermore, both sides agree that the deleterious side effects of technological change can not be ignored. Finally, they both admit that advanced automation is a fact of life, and as such is deserving of serious evaluation.

Of course, the similarities in the two positions are interesting in and of themselves. But to the extent that they remind us that even in the midst of a heated public debate there are usually common grounds for cooperation, they may have far greater significance. If cooperation has not been the rule thus far, then it is equally true that cooperative efforts have not been completely avoided. The key is to find more ways to bring together individuals from both camps who represent industry, government, academe, the media, and the public in an effort to extend the common grounds and to seek a responsible policy on technology dissemination.

In a society that practices adversarial politics daily, cooperation does not come easy. But it may very well be preferable to the alternatives suggested by history. Automation advocates in particular should recall that the federal government first entered into the business of regulating industry in response to public outcry over the human consequences of poorly conceived technology -- namely steamboat boilers.[69] And critics of automation might do well to recall that public condemnation of automation during the 1950's and early 1960's did not prevent the present predicament. Indeed, it may have played a role in creating it.[70] If the experts on both sides pursue their present course of media confrontation, then history suggests that our ability to attain the common goals of industrial revitalization and social justice is threatened. If, on the other hand, the experts work together for these common goals, then we can look to the future with anticipation and even enthusiasm.

NOTES

1. For discussions of the growing influence of high technology, especially advanced automation, on all aspects of life, see: Cornish, Blake M., "The Smart Machines, Implications for Society of Tomorrow," The Futurist, pp. 5-13; (August 1981); Edward Cornish, "The Coming of an Information Society", The Futurist pp. 14-21 (April 1981); Also see: OMNI (April 1983). This entire issue is devoted to robots and society.

2. Statement by James Baker, Vice-President of General Electric. Taken from the following: Rita R. Schreiber, "Robotics in the Eighties", Robotics Today, pp. 39-40, (October 1982).

3. The slogan, "Automate, immigrate, or evaporate", has become a popular slogan with industry during the 1970s and 1980s. See: Schreiber, pg. 39. (Note 2).

4. Relative costs and efficiency figures were provided by Unimation, Inc. Samuel Walters, "The Robots Are Coming", in Mechanical Engineering, pg. 102 (December 1980); Jeffrey E. Christian, "Putting the Robot to Work", Robotics Today pp. 36-37, (Spring 1981).

5. For example, in an article devoted to flexible automation, Gene Bylinsky presents figures for the fully automated factory of the Yamazaki Machine Works, Ltd., to support his general claims about the new technology. With a reduction in the work force of 94 percent, and with a reduction in production time for a given order from three months to three days, Yamazaki believes that in five years of operation the firm's after tax profits will be $12 million, compared with just $800,000 using the previous system.

See: Gene Bylinsky, "The Race to the Automated Factory", Fortune, Volume 107, pp. 52-61 (February 21, 1983).

6. "Burlington Uses Shorter Product Runs as it Makes Big Leap in Modernization", The News and Observer, pg. 3D, (March 26, 1985).

7. Ward, Bernie, "Robotics and Reality", Sky, pp. 23-30, (November 1982).

8. M. Eugene Merchant, "One Man's View", The Industrial Robot, pg. 151 (September 1980).

9. Dan T. Moore, Jr., "Will Robots Save Democracy?", The Futurist, 14-9, (August 1981).

10. Carlson, Eugene, "Michigan Sees Robot-Making as a Solution to Its Troubles", The Wall Street Journal, pg. 37, (March 23, 1982).

11. Bagamery, Anne, "No Policy is Good Policy", Forbes, pg. 140, (June 18, 1984).

12. Polcyn, Stanley, "Robots and Our Jobs", Omni, pg. 35, (April 1983).

13. Christian, pp. 36-37. (Note 4).

14. Bartlet, Bruce, "The Luddite Answer to Unemployment", "The Wall Street Journal, pg. 18, July 18, 1983.

15. Statements claiming that the unions support the pro-automation argument can be found in the following: "U.S. Unions - An Englightened View of Robotics", "The Industrial Robot, pp. 228-30 (December 1980); "Unions Take a Positive Attitude to the Introduction of Robots", in The

Industrial Robot, pp. 228-29.

16. The complete statement, made during the 1982 annual conference of the TYAA, was reported in Textile Industries, pg. 58, (October 1982). "There are processes that involve tasks that people can't do, or won't do, or shouldn't do (such as creeling and doffing texturing machines with jumbo yarn packages), and these jobs are best done by robots."

17. Eugene Merchant relies heavily on Daniel Bell's work to document the transition already underway to post-industrialism. See: Daniel Bell, The Coming of the Post-Industrial Society; A Venture in Social Forecasting New York; Basic Books, Inc., pg. 156, (1973). For the quotation by Merchant, see: Merchant, "One Man's View", The Industrial Robot, pg. 151, (September 1980).

18. For discussions concerning the shortage of engineers and skilled technicians, see: Smith Rathmill, "One Man's View", The Industrial Robot, pg. 143, (September 1981); "Shortage of Engineers, Computer Experts Seen for High-Tech Industry", The Raleigh Times, pg. 19A, (June 30, 1983); Michael Flagg, "Lack of Trained Workers Holding Robots Back in N.C.", The News and Observer, pp. 12-13D, (January 27, 1985); Robert R. Trouteaud, "What the Robot Revolution Needs: Smarter People", Robotics Today, pg. 42-43, (June 1982).

19. Stauffer, Robert N., "Meeting the Challenge of Robotics Training", Robotics Today, pp. 32-34, (June 1982).

20. For an excellent discussion of the advantages of in-house training for skilled personnel, see: Lloyd R. Carrico, "Robotics Training: A Comprehensive Plan", Robotics Today, pp. 38-39, (June 1982). Mr. Carrico is the Super-

visor of Training, Industrial Robot Division, Cincinnati Milacron, Inc.

21. Schreiber, Rita R., "Recognizing Competence in Robotics", Robotics Today, pp. 55-56, (April 1982).

22. For discussions of high-technology research in American Universities, see: Amram, Fred M., "Robotics: The Human Touch", Robotics Today, pg. 28, (April 1982); Ward, pp. 23-30 (Note 7).

23. Ward, pp. 23-30 (Note 7).

24. Botkin, James, Dan Dimancescu, and Ray State, Global Stakes: The Future of High Technology in America, pg. 154, (Cambridge: Ballinger Publishing Co., 1982).

25. For discussions of industry-university collaboration, see: Amram, 28 (Note 22); Rathmill, 143 (Note 18); "Government Report Praises NCSU for its Role in Forging University-Industrial Relations", The Journal of North Carolina State University, Volume 15, pg. 4, (January 1984); "Robotics Institute Teams Development in University and Industry", The Industrial Robot, pp. 246-247, (December 1981).

26. "New Robot at NCSU", The Leader, pg. 7, (May 5-12, 1983).

27. "Government Report Praises NCSU. . .", The Journal, pg. 4 (Note 25).

28. "'Automate, Integrate -- or Evaporate', N.C. Firms Told," The News and Observer, pp. 12-13D, (January 27, 1985).

29. Cosineau, David T., "Robots Are Easy. . . It's Everything Else That's Hard", "The Industrial Robot, pp. 50-56,

(March 1981).

30. Winter, Ralph E., "Streamlining Firms May Not Bring the Big Benefits That Some Expect", The Wall Street Journal, pg. 33, (October 25, 1982).

31. For a discussion of North Carolina State University's new Integrated Manufacturing Systems Engineering Institute, which is designed to "help industry increase production of goods with higher quality and lower costs, see: "Engineering Institute to Aid Manufacturing", The Journal of North Carolina State University, Volume 15, pg. 1, (January 1984). For a discussion of industry programs, see: Laura Landro, "GE Promotes Factory Automation, But Some Doubt Big Market Exists", The Wall Street Journal, pg. 31, (October 21, 1982); "Spotlight on Production Engineering Research Association", The Industrial Robot, pp. 214-216, (December 1980).

32. Landro, pg. 31 (Note 31).

33. This World of Work Report statement is quoted in the following: Neale W. Clapp, "Three Laws of Industrial Robotics", Robotics Today, pp. 20-22, (Spring 1980).

34. This statement is taken from: Gennaro C. Macri, "Analysis of First UTD Installation Failure", Society of Mechanical Engineers Technical Paper MS77-735, 1977. It is quoted in: Clapp, "Three Laws of Industrial Robots", pg. 20 (Note 33).

35. Clapp, Neale W., "Management Resistance to Robots", Robotics Today, pp. 39-41, (Spring 1981).

36. For a discussion of the danger of worker resistance, see: Clapp, "Three Laws of Industrial Robots", pg. 19 (Note 33).

37. Gwynne, Peter, "Robot Revolution", U.S. Air, pg. 95, (June 1983).

38. Statement by Donald Dominick can be found in the following: Schreiber, "Robotics in the Eighties", pg. 42 (Note 2).

39. For a discussion of the importance of top-management involvement in retraining programs which use the Japanese as a model, see: Kenichi Ohmae, "Steel Collar Workers: The Lessons from Japan", The Wall Street Journal, pg. 26, (February 16, 1982).

40. Awerman, Jackie and Diana Cappello, "Positive Employee Relations Paves the Way for Robots", Robotics Today, pp. 35-36, (December 1981).

41. "Burlington CEO Updates Automated Apparel Project", Textile World, pp. 35-38, (February 1985).

42. Simon, Herbert A., "Prometheos or Pandora: The Influence of Automation on Society", IEEE Spectrum Volume 14, pg. 71, (November 1981).

43. "Engineering Practice in an Era of Social Change -- An Interview with Martin Goland, Mechanical Engineering Volume 104, pp. 46-48, (May 1982).

44. The Ayres and Miller study was quoted in: Fred Guterl, "An Unanswered Question: Automation's Effect on Society", IEEE Spectrum Volume 20, pg. 90, (May 1983).

45. Guterl, Fred, "An Unanswered Question: Automation's Effect on Society", IEEE Spectrum, Volume 20, pp. 89-90, (May 1983).

46. Brooks, Geraldine, "Despite the Recovery, Some Lose Jobs

Now and it's a Hard Blow", The Wall Street Journal, pp. 1 & 12, (August 5, 1983).

47. Statement by James Phelps, Hewlett-Packard Group Personnel Manager. Taken from: Bob Kuttner, "The Declining Middle", The Atlantic Monthly, pg. 66, (July 1983).

48. Argote, Linda, Paul S. Goodman, and David Schkade, "The Human Side of Robotics: Results from a Prototype Study on How Workers React to a Robot", Robotics Institute Technical Paper CMU-RI-TR-83-11, pp. 7-10, (Carnegie-Mellon University, 1983).

49. Shaiken, Harley, "Automation in Industry: Bleaching the Blue Collar", IEEE Spectrum, Volume 21, pp. 77-79, (June 1984).

50. This quotation by Harley Shaiken was taken from: Kathleen Stein, "Robo-Shock!", OMNI, pg. 46, (April 1983).

51. Bluestone, Barry and Bennett Harrison, The Deindustrialization of America: Plant Closings, Community Abandonment and the Dismantling of Basic Industry, (New York: Basic Books, Inc., 1982).

52. Levin, Doron P., "Shift from Middle Class to Poverty Proves Painful for Longtime Jobless", The Wall Street Journal, pg. 21, (June 25, 1982). Also see: Joann S. Lublin, "How to Help Displaced Workers?", The Wall Street Journal, pg. 16, (July 29, 1983).

53. Ayres, Robert and Steve Miller, "The Impact of Industrial Robots", Robotics Institute Report", Carnegie-Mellon University, November 1981.

54. Some critics seem to feel that union attitudes and

activities actually contribute to the perception that automation-induced unemployment does not represent a serious social problem. By focusing on "take-aways", such as personal absence days, bonus time, and cost-of-living increases, the unions have, until very recently neglected the automation-related issues of job security and retraining. See: Guterl, pg. 92 (Note 45).

55. Simison, Robert L., "Ford, UAW Appear Headed for Accord on Senior Workers' Guaranteed Income", The Wall Street Journal, pg. 2, (February 12, 1982).

56. Rose, Frank, "The Mass Production of Engineers", Esquire, pg. 74-84, (May 1983).

57. Peters, Thomas J., and Robert H. Waterman, Jr., In Search of Excellence: Lessons From America's Best-Run Companies, pp. 238-39, (New York: Harper & Row, Publishers, 1981).

58. U.S. General Accounting Office, "Advances in Automation Prompt Concern Over Increased U.S. Unemployment", GAZ/AFMD-82-44, pg. 15, (May 1982).

59. Klopman, William A., Jr., "Are Textile Jobs Expendable?", Modern Textiles, volume 58, pp. 24-25, (November 1977).

60. Brecher, John, with Alexander Stille, "'Telescabbing': The New Union Buster", Newsweek, Volume 102, pp. 53-54, (August 29, 1983).

61. Kuttner, Bob, "The Declining Middle", The Atlantic Monthly, pp. 60-72, (July 1983).

62. Lublin, Joann S., "How to Help Displaced Workers?", The Wall Street Journal, pg. 16, (July 29, 1983).

63. "Judge Orders Up to $124 Million Outlays for Import Aid and Training of Jobless", The Wall Street Journal, pg. 2, (August 16, 1983).

64. For discussions of the automation of office work, see: "Diebold's Shift to Automated Teller Works", The Wall Street Journal, pg. 33, (July 15, 1982).

65. "Firms Start Using Computers to Take the Place of Salesmen", The Wall Street Journal, pg. 33, (July 15, 1982).

66. Wolin, Merle Linda, "Dirty Work: Americans Turn Down Many Jobs Vacated by Ouster of Aliens", The Wall Street Journal, pg. 1, (December 6, 1982).

67. Pynchon, Thomas, "Is it O.K. to be a Luddite?", The New York Times Book Review, pg. 41, (October 28, 1984).

68. Noble, David F., Forces of Production: A Social History of Machine Tool Automation, (New York, Knopf, 1984).

69. Burke, John G., "Technology and Government", Technology and Social Change in America, pp. 99-119, (Editor: Edwin P. Layton, Jr.).

70. Bonham, Julia C., "Assessing the Human Impact of Automation: The Prolonged Public Debate", Unpublished manuscript, 1984.

Other Noyes Publications

TEXTILE IDENTIFICATION, CONSERVATION, AND PRESERVATION

by

Rosalie Rosso King
Western Washington University

Textile identification, preservation and conservation methods, as presented in this book, include both simple and complex processes that relate to techniques used in both the arts and sciences. The process of identification and conservation must be exacting to insure that mistakes are not made which would impair the beauty or lasting quality of the textile piece. An in-depth understanding of fibers, assemblages of fibers, and fabrics, plus a study of their reaction to cleaning procedures, is necessary to insure that future generations will have the opportunity to enjoy textile pieces currently being produced as well as those collected from the past.

Because each culture has developed textile artifacts, the world is rich in a diversity of pieces and an abundance of soft goods. Yet some items were made in very limited quantities for extremely specialized end uses. Some have required years of often painful labor for their construction. The primary aim in the writing of this book is to provide identification, cleaning and conservation techniques for all textile fibers and fabrics found today. A review of textile and related research findings through the early 1980s has been included, as the importance of clean, well-cared-for textiles cannot be overlooked in relation to the management of a collection, museum storage, or to the expected life span of actual pieces. However, some textile pieces containing historically important soil or foreign matter, or those that would be altered or changed, should be left untouched. These cases are also discussed.

Because of their increased use in recent years, man-made fibers and fabrics are given detailed coverage in the book. The condensed table of contents given below lists **chapter titles and selected subtitles.**

ISBN 0-8155-1033-0 (1985) **356 pages**

Other Noyes Publications

TEXTILE MARKETING MANAGEMENT

by

Gordon A. Berkstresser III
North Carolina State University
School of Textiles

This book focuses on current textile marketing practices in the U.S. and explores the modern basis for those practices. The movement toward an international textile marketing arena is fully discussed.

Case studies taken from real situations are included in most chapters to provide realistic examples of the problems with which textile marketing managers must deal. The book will be useful to people in the industry as well as to students of textiles and related subjects.

A condensed table of contents listing **chapter titles and selected subtitles** is given below.

ISBN 0-8155-0975-8 (1984) **$32** **135 pages**

Other Noyes Publications

FABRIC FORMING SYSTEMS

by

Peter Schwartz
Cornell University

Trevor Rhodes
Levi Strauss & Co.

Mansour Mohamed
North Carolina State University

This basic technical text on fabrics and fabric formation highlights both woven and knitted fabrics, as well as the major categories of nonconventional fabrics. The main fabric forming systems, and the structures produced, in weaving, knitting, and nonconventional fabric production are detailed.

The book covers the preparation of yarn for weaving and knitting, loom mechanisms, classification of looms, woven fabric design, weft knitting and weft knit structures, warp knitting and warp knit structures. Included also is a brief introduction to nonwoven fabric web formation and different bonding or manufacturing systems such as saturation bonding, print bonding, spunbonding, needle punching and stitch bonding. Tufting and flocking are also discussed.

The book will be particularly suitable for textile managers, students, or professionals of nontextile background seeking an overview of textile manufacturing.

Detailed diagrams illustrate the written text throughout the book. A table of contents listing **chapter titles and selected subtitles** is given below.

ISBN 0-8155-0908-1 (1982) **175 pages**